Studying Primates

How to Design, Conduct and Report Primatological Research

Primatology draws on theory and methods from diverse fields, including anatomy, anthropology, biology, ecology, medicine, psychology, veterinary sciences, and zoology. The more than 500 species of primate range from tiny mouse lemurs to huge gorillas, and primatologists collect data in a variety of environments including field sites, research facilities, museums, sanctuaries and zoos as well as from the literature. The variability in our research interests, study animals and research sites means that there are no standard protocols for how to study primates. Nevertheless, asking good questions and designing appropriate studies to answer them are vital if we are to produce high-quality science. This accessible guide for graduate students and postdoctoral researchers explains how to develop a research question, formulate testable hypotheses and predictions, design and conduct a project, and report the results. The focus is on research integrity and ethics throughout, and the book provides practical advice on overcoming common difficulties researchers face.

JOANNA M. SETCHELL is Professor of Evolutionary Anthropology at Durham University, UK, and Editor-in-Chief of the *International Journal of Primatology*. She is an international expert in primatology with an extensive publication record. She is currently President of the Primate Society of Great Britain and has served as Vice-President for Research of the International Primatological Society. She is co-editor of *Field and Laboratory Methods for Primatologists* (Cambridge University Press, 2011).

Studying Primates

*How to Design, Conduct
and Report Primatological
Research*

JOANNA
M. SETCHELL
Durham University

CAMBRIDGE
UNIVERSITY PRESS

CAMBRIDGE
UNIVERSITY PRESS

University Printing House, Cambridge CB2 8BS, United Kingdom

One Liberty Plaza, 20th Floor, New York, NY 10006, USA

477 Williamstown Road, Port Melbourne, VIC 3207, Australia

314–321, 3rd Floor, Plot 3, Splendor Forum, Jasola District Centre, New Delhi – 110025, India

79 Anson Road, #06–04/06, Singapore 079906

Cambridge University Press is part of the University of Cambridge.

It furthers the University's mission by disseminating knowledge in the pursuit of education, learning, and research at the highest international levels of excellence.

www.cambridge.org
Information on this title: www.cambridge.org/9781108421713
DOI: 10.1017/9781108368513

First published 2019

Printed in the United Kingdom by TJ International Ltd, Padstow Cornwall

A catalogue record for this publication is available from the British Library.

ISBN 978-1-108-42171-3 Hardback
ISBN 978-1-108-43427-0 Paperback

Cambridge University Press has no responsibility for the persistence or accuracy of URLs for external or third-party internet websites referred to in this publication and does not guarantee that any content on such websites is, or will remain, accurate or appropriate.

To the next generation of primatologists

Contents

Preface

There are more than 500 species of primate, ranging in size from the ~30 g Madame Berthe's mouse lemur to the ~250 kg male Grauer's gorilla. Some species are arboreal, others terrestrial. Some live in large groups, others forage alone. Some species are diurnal, some nocturnal, and still others are active at any time of day or night. Primatology draws on theory and methods from diverse fields, including anatomy, anthropology, biology, ecology, medicine, psychology, veterinary sciences, and zoology. Primatologists collect data in field environments with various levels of anthropogenic influence, animal sanctuaries, research facilities, museums and zoos, and from the literature. We are motivated by discovery, conservation, and welfare. The variability in our study animals, research interests, and motivations means that there are no standard protocols or off-the-shelf recipes for how to study primates. Nevertheless, asking the right questions and designing appropriate studies to answer them are vital if we are to produce high-quality science. These are also the topics that students often find most difficult.

This book guides readers in how to think scientifically. It covers the skills needed to assess published studies critically, develop a research question, formulate testable hypotheses and predictions, design and conduct a study that will test the predictions, select appropriate measures and samples, analyse the data, interpret the results, draw conclusions about the data in relation to the original question, and report the results. I emphasise ethics and the need for honest, rigorous, and transparent science throughout. I explain common misconceptions and problems in primatology and how to resolve them. I include the difficulties researchers can face, including imposter syndrome, field-work in difficult environments, and reverse culture shock. I highlight **key terms** the first time I use them and use text boxes to cover issues in more depth than in the main text. Each chapter includes a summary and suggestions for further reading.

This book will be valuable for postgraduate and postdoctoral scholars in primatology and allied disciplines. It will also be useful for final-year undergraduates contemplating their dissertation and for those who teach undergraduates, advise postgraduates, mentor postdoctoral scholars, and conduct their own research.

Author royalties will be donated to the International Primatological Society Conservation fund.

Acknowledgements

This book distils what I've learnt over more than 20 years. It is based on my experience in my own research, editing the *International Journal of Primatology*, collaborating with students, and discussions with colleagues and friends. I don't claim to be a perfect scientist and this book reflects what I've learnt through what I've done, rather than what I've done. Informally, I call it *How not to mess it up*, with a good helping of *Don't be a jerk*. I hope that it will help people to do good research in primatology.

The best collaborator is a friend you do science with. Thank you to my many friend-collaborators, particularly Kristin Abbott, Sandra Bell, Robin Bernstein, Marie Charpentier, Elena Cunningham, Debbie Curtis, Wendy Dirks, Phyllis Lee, Barthélémy Ngoubangoye, Klara Petrzelkova, and Steve Unwin. Charlie Lockwood, it was a privilege to know you.

I have learnt a huge amount from editors, reviewers, and authors at the *International Journal of Primatology*, and from colleagues in the Primate Society of Great Britain and the International Primatological Society. Thank you all.

I am very grateful for the generous support of the Centre International de Recherches Médicale, Franceville, Gabon, and all the staff of the Centre du Primatologie during our long collaboration. In addition to generous collaboration and great friendship, CIRMF have always provided rent-free accommodation.

Many people and institutions helped me as a PhD student and postdoctoral scholar. Alan Dixson supervised my PhD at Madingley, the School of Life Sciences at Roehampton University gave me free rein as a postdoctoral fellow, and Leslie Knapp was my postdoctoral mentor in Biological Anthropology at Cambridge. E. Jean Wickings introduced me to mandrills, loaned me a car, and was an exceptionally generous host in Franceville, Gabon. I stayed with Jason Brown rent-free when my PhD stipend ran out. Benoît Goossens, Marc Ancrenaz, and friends gave me the opportunity to experience life in Sabah, Malaysia, and the Institute of Tropical Biology and Conservation, Universiti Malaysia Sabah, hosted me as a guest lecturer. Chloë Aldam and Chris Bell let me sleep in their living room and house-sit when I was home from the field as a postdoctoral scholar. Corinne and Bernie Curtis charged me far less than the going rate for rent for their flat in London. Thank you all.

Thank you to Fanny Cornejo and Cesar F. Flores-Negron for lengthy discussions of how to teach the scientific method in the inspirational environment surrounding Cocha Cashu Biological Station, and to our students on the 1er Curso de Campo para el Estudio de la Ecología, Comportamiento, y Conservación de Primates, who put up with my abysmal Spanish. Thank you to all the people who have guided me to see wild primates.

I thank Helen Ball at Durham University for encouraging me to write this book. Thank you to the research students and postdoctoral scholars I have collaborated with, whose experiences have contributed directly to this book: Esther Clarke, Ben Coleman, Emilie Fairet, Katharine Flach, Ingrid Grueso-Dominguez, Steffi Henkel, Caroline Howlett, Dhana Hughes, Sharon Kessler, Simone Lemmers, Pedro Mendez-Carvajal, Gaby Mendoza Nakano, Lucy Millington, Rodrigo Moro-Rios, Danson K. Mwangiri, Marie-Claire Pagano, Kat Shutt, Pete Tomlin, Stefano Vaglio, Sian Waters, and Miles Woodruff.

For constructive comments on the proposal for this book, I thank Trudi Buck, Sarah Elton, Russell Hill, Rachel Kendal, Stuart Semple, and four anonymous reviewers. For help with specific points or feedback on individual sections and chapters, I thank Alec Ayers, Helen Ball, Rob Barton, Trudi Buck, Sarah Elton, Jeremy Kendal, Rachel Kendal, Simone Lemmers, Kesson Magid, Gaby Mendoza Nakano, Danson K. Mwangiri, and Chris Schmitt. For feedback on multiple chapters, I am very grateful to Roger Mundry, who reviewed an early version of the stats content. Sharon Kessler, Bruce Rawlings and Pete Tomlin provided invaluable feedback on the first full draft. Celia Deane-Drummond provided a very helpful external eye and advice on ethical issues. Trudi Buck and Sarah Elton provided encouragement and very helpful feedback on the second draft. Andrew Mein provided very helpful feedback on the index and the proofs. I am grateful to Dietmar Zinner and Christian Roos for creating Figure 7.1. If I've missed anyone, I apologise. All remaining errors are, of course, mine.

Thank you to Megan Kiernan of the life sciences editorial team at Cambridge University Press for her encouragement when I told her about the idea for this book, and for obtaining reviews and approval for the proposal. Thank you to Aleksandra Serocka and Ilaria Tassistro for their patience in extending the word count and the deadline and to Annie Toynbee for guiding me through the publication process. I am also grateful to Ishwarya Mathavan and Anbumani Selvam for copy-editing the manuscript.

Thank you to my friends, particularly Gill Atkinson, Trudi Buck, Viv Kent, Liz Pavey, and Jane Shuttleworth, for putting up with my obsession with this book and for their support.

Merci à ma famille gabonaise pour une hospitalité sans pareil: Hanny Marva Ipendangoye et ta famille, Adiss Adebo, Macair et Katy Leyama, Rodrigues (Guesoh) Louembet, Clementine Matha, Rainey (Matcho) Mombet, Yves-Landry (Ya Mouk) Mouketou, Jean-François (Adjo) Mouori, Aristide Nguembi, Peckoss Nkoghe, et tous les enfants. Sosthène John, Aymar Mangoumba, et Bettina Salle, vous nous avez quitté trop tôt.

Christian Nadjimbaye, tu es une grosse partie de ma force, malgré la distance.

And, finally, thank you to my parents, Michael and Mary, for their support throughout my career, despite their worry that I would never get a proper job. I did, and this book is what I have learnt.

1

Asking Questions about Primates

All observation must be for or against some view, if it is to be of any service
Charles Darwin[1]

Like all science, studying primates is about asking the right questions in the right way. Most studies of primates fall within the life sciences, so I focus on the scientific method in this book.[2]

In this chapter I first introduce how science works, then look at what it takes to be a primatologist. Finally, I outline the contents of the rest of the book.

1.1 HOW SCIENCE WORKS

All cultures produce and accumulate knowledge to understand and explain the natural world. Science is one such knowledge system.

Science is an attempt to explain observations of natural phenomena such that we can predict future observations. We begin with observations, either our own or those reported in the literature. We look for patterns in those observations and propose explanations for the patterns we identify (**hypotheses**). We then use those hypotheses to derive **predictions** about what will happen under specific circumstances, and collect, analyse and interpret new observations (**empirical data**) to test whether our predictions are upheld. New data may support a hypothesis, help to refine it, help to refute it or inspire us to suggest a new hypothesis. We retain or refine hypotheses that successfully predict observations, but we cannot prove them (Box 1.1). We reject hypotheses if they are not consistent with our observations.

---

[1] Darwin Correspondence Project, 'Letter no. 3257'. www.darwinproject.ac.uk/DCP-LETT-3257 [Accessed 8 April 2017].

[2] Studies of human relations with primates need a broader approach, including the social sciences.

Box 1.1: Common misconceptions about science

Science is often misrepresented in the media, leading to misunderstandings about how science works. Common misconceptions include the following points.

Science Is about Facts
Science is about the process of asking questions, not about facts. In other words, it is about what we don't know, not what we do know.

Scientists Prove Hypotheses
We retain or refine hypotheses that successfully predict observations; we do not (and cannot) prove them. All scientific knowledge is provisional, although some hypotheses are very strongly supported.

Scientists Seek Evidence That Supports an Explanation
Good scientists look for evidence that their ideas are wrong. We don't seek evidence that supports an explanation.

Science Is about Breakthroughs
Although media coverage often hypes breakthroughs (as well as bizarre and scary stories), science is usually incremental and major breakthroughs are rare.

We Can Achieve a Complete Understanding of a Question
We rarely provide a definitive or simple answer to a question, and usually end with more questions than we began with at the end of a study. We never achieve a complete understanding of a question because each new answer opens further questions.

Using Individual Cases to Counter General Models
Media coverage and non-scientists often highlight an individual case that forms an exception to criticise a study. However, scientific hypotheses are simplified models of complex real-world phenomena, and they do not explain every detail, every situation or every case, so an exception does not necessarily mean that the model is false.

There Is a Single Explanation for a Phenomenon
People also often confuse mechanistic (how things work) and functional (why they happen) explanations for a phenomenon, putting one forward as a counterargument to the other. However,

Box 1.1: (cont.)

the two explanations do not compete, and we need to understand both to understand a phenomenon.

Science Is Completely Objective
Scientists strive to be objective, but we are human, and subject to bias.

Confusing Statistical Significance with Real-World Importance
Scientists use the word *significant* to mean that results are statistically significant. This is often misinterpreted as *meaningful* in terms of the real world. However, statistical significance does not measure the size or importance of an effect. The size of the *difference* is more important. We'll look at this in more detail in Chapter 5.

FURTHER READING

Goldacre B. 2009. *Bad Science*. London: Fourth Estate. An entertaining review of how the media misunderstands and misrepresents health research.

A good scientific model explains as many observed patterns as possible. We use **Occam's razor** (also known as the law of **parsimony**) as a guide. Occam's razor holds that we should retain the simplest hypothesis – the one that makes the fewest assumptions. This guards against adding further explanations (*ad hoc* **hypotheses**) to reinforce a favoured but unsupported model to explain patterns the original model failed to predict.

Models simplify reality and explain general patterns. They do not explain all individual observations. For example, if males are, in general, more aggressive than females, that doesn't mean that *all* males are more aggressive than *all* females. The variation among individuals is the raw material of evolution by natural selection. Similarly, a pattern found in one species does not necessarily apply (**generalise**) to other species. In other words, there is no *typical* primate, either within or across species.

Models explain natural phenomena at **different levels** of analysis, resulting in very different answers to the same question. A **proximate explanation** is an immediate cause of an observed phenomenon. Proximate explanations answer *how* questions. An **ultimate explanation** addresses how a trait contributes to an individual's ability to propagate its genes. Ultimate explanations answer *why* questions. These different levels of analysis are distinct and equally valid. They complement one another, rather than being competing alternatives, but are frequently confused. We need to address both to understand a phenomenon.

Often, we cannot observe or measure phenomena directly, so we infer an explanation based on **indirect evidence**. For example, we cannot measure an organism's mental state, so we might use observations of behaviour to infer it. Other phenomena are hard to observe because they are rare or cryptic, requiring creative solutions if we are to study them.

Over time, we accumulate and assess evidence and evaluate whether it supports or falsifies our models. A single study can't give a definitive answer to a question. Individual studies are often described as bricks in a wall – a useful if simplistic metaphor. We test findings to determine whether we can reproduce the original results using the same data (**reproduction**), and whether independent investigations (i.e. new data) will produce the same findings (**replication**).

Replication can be conducted at different levels, with a trade-off between establishing the validity of findings and their generality (Box 1.2). Repetition makes science self-correcting. If multiple lines of evidence support the same hypothesis, then we can be more confident in that explanation. Convergence of evidence from multiple approaches, each with different assumptions and sources of potential bias, is termed **triangulation** or **consilience**, and leads to robust conclusions. Break-throughs, or **paradigm shifts**, occur when new observations, ideas, findings, or methods alter how we view a problem. The metaphorical wall falls down and is rebuilt in a different form.

Hypotheses that can be disproven (**falsified**) and tests that exclude alternative hypotheses (**strong inference**) were the dominant philoso-phies of twentieth-century science. However, many questions in primat-ology are better expressed in terms of the size of an effect, or the strength of a relationship, rather than a binary yes/no as to whether the effect or relationship exists.

1.2 WHAT IT TAKES TO BE A PRIMATOLOGIST

Primatologists study primates to understand them, conserve them and promote their welfare. We draw on diverse theory and methods from disciplines including anatomy, anthropology, biology, ecology, medi-cine, psychology, veterinary sciences and zoology. We work in labora-tories, museums and libraries, and at animal sanctuaries, captive

Box 1.2: Types of replication

Replication is essential to scientific progress. It tells us whether the original results are reliable and meaningful.

An **exact replication** (or **direct replication**) aims to duplicate the methods of an earlier study completely. However, this is often impossible in primatology, where the exact conditions are unique to a study. **Partial replication** ranges from **close replication**, using the same methods to replicate the original study as far as possible, to **conceptual replication**, which evaluates the same hypothesis as the original study, but using different methods. Close replication is an excellent test of whether the results are repeatable and reliable, but not of whether they apply in other settings (**generality** or **generalisability**). In contrast, conceptual replication is a poor test of validity, but a better test of generality. **Quasi-replications** expand the scope of study to a new species or system. They are not effective at testing validity but are useful to assess the generalisability of findings across species.

The terms *reproducible* and *replicable* are often used interchangeably but have distinct meanings in science. Reproducible means that if we use the same dataset and methods we will obtain the same statistical results as those reported. Replicable means that an independent test of the same hypothesis, using the same methods, will obtain the same findings.

FURTHER READING

Kelly CD. 2006. Replicating empirical research in behavioral ecology: How and why it should be done but rarely ever is. *The Quarterly Review of Biology* 81: 221–236. https://doi.org/10.1086/506236. Explains why we need to replicate empirical studies and why we often don't.
Munafò MR, Smith GD. 2018. Repeating experiments is not enough. *Nature* 553: 399–401. https://doi.org/10.1038/d41586–018-01023-3. A commentary on the need for triangulation.
Nakagawa S, Parker TH. 2015. Replicating research in ecology and evolution: feasibility, incentives, and the cost-benefit conundrum. *BMC Biology* 13: 88. https://doi.org/10.1186/s12915–015-0196-3. Advocates replication in ecology and evolution, explaining the various types of replication.

research facilities, zoological parks (zoos) and field sites. We develop theory, collate and analyse existing data, create computer models, study fossils, observe behaviour, conduct experiments and do laboratory analyses. These diverse areas all require a combination of deep familiarity with our study animals, a strong grasp of theory, an excellent understanding of study design and statistical analysis, careful thought and planning, and the ability to communicate our research to academics and the general public.

Deep familiarity with our study animals and sharply honed observation skills are essential to a project. We can't design a project and we can't interpret our findings without understanding how our study animals behave. Observing other species can also provide illuminating comparisons and lead us to pose new questions.

A strong grasp of current theory is essential to put our observations into context and recognise interesting research questions. Simple descriptive research provides important **natural history** information and is essential to generate new hypotheses, but it is impossible to identify interesting questions without a sound understanding of theory.

Careful thought about study design and how to analyse our data is crucial to the success of a project. It can be very tempting to jump straight into data collection, using familiar methods or copying other researchers, without stopping to think about how we intend to analyse our data to address a specific question. However, data can be useless if the methods were not designed to answer a question or the number of cases we study (the **sample size**) is inadequate to draw any general conclusions. Moreover, mining data we have already collected for patterns runs a serious risk of testing hypotheses suggested by the data (more on what's wrong with this in Chapter 2). These are common, but preventable, errors.

The ability to communicate our ideas to readers is an indispensable component of research. Some of us are native English speakers, but none of us are native scientific English speakers, so we all need to learn this. We must also learn the skills to communicate with a broader audience.

Developing these diverse skills takes practice. We must be prepared to accept and act on critical feedback to refine and improve our work, however hard this may be. Science can be daunting and it's normal to feel overwhelmed at times (Box 1.3). Many scientists suffer from the feeling that they don't belong or might be exposed as a fraud. Box 1.4 explores the sources of this **imposter syndrome**, and how to handle it.

Beyond this, being a primatologist requires passion, creativity, adaptive perfectionism, curiosity, tenacity, creativity, stamina, resilience, flexibility, patience, commitment, attention to detail, openness

Box 1.3: Coping when you're overwhelmed

Science can be hard. At times your to-do list can seem unending. It's normal to feel like you're making a mess of things at times, and for tasks to take much longer than you had planned. Moreover, health, financial, and personal issues often intervene in a project. It can seem as though students and researchers are expected to have no other concerns in life and to devote all their time to research. However, this isn't realistic. Researchers are human, with responsibilities, financial concerns, caring responsibilities, families, hobbies and so on.

When you feel overwhelmed, take a step back and look at the bigger picture. Don't just dive into the first task you have to hand. Instead, list your goals and break them into tasks. Break tasks into sub-tasks until you reach small, manageable tasks. Prioritise among those tasks and put them into a timeline. Distinguish between urgent and important tasks and ensure what you're doing moves you towards your goals. Remember that you don't have to do everything, just something. Monitor your progress and readjust your timeline based on your experience. Reward yourself for achieving goals and be kind to yourself if you don't. Figure out what works best for you (e.g. where and what time of day you work best).

Take breaks. You can work flat-out for a while, but you can't sustain that for any length of time without negative effects on you and on your work. Sleep, eat regularly and well, exercise and take breaks from work. Spending time with people who are not researchers is a good way to take a break.

Scientific writing groups, either in person or online, can help with support, and strategies to combat **procrastination** (postponing tasks that you need to accomplish), or make it productive.

If something knocks you off course, remember that that is normal. Assess what happened. Many students and researchers experience anxiety and depression. If you do, you may feel alone, but you are not. Tell someone you trust, so that they can help. Seek professional help. If you're attached to an institution, find out what support measures are available. Awareness of mental health issues among postgraduate students and academics is improving, but attitudes to mental health vary and you may be treated unfairly if you disclose ill health. Many supervisors, advisors and mentors are

Box 1.3: (cont.)

supportive, but some are not. Know your rights and document any bias you experience. Online networks for students and researchers with mental health concerns are a good source of support.

If you need to take a break from your research due to ill health, you may not be able to jump back into work at 100%. Chronic health conditions may mean that you need to manage your work carefully. Be honest with yourself and with your co-workers about how much you can achieve. Again, you may face discrimination.

FURTHER READING

Evans TM, Bira L, Gastelim JB, Weiss LT, Vadnerford NL. 2018. Evidence for a mental health crisis in graduate education. *Nature Biotechnology* 36: 282–284. https://doi.org/10.1038/nbt.4089. Reports the results of a survey highlighting the serious problem with mental health in graduate students.

Guthrie S, Lichten CA, Van Belle J, Ball S, Knack A, Hofman J. 2017. *Understanding Mental Health in the Research Environment: A Rapid Evidence Assessment.* Santa Monica, CA: RAND Corporation. www.rand.org/pubs/research_reports/RR2022.html [Accessed 2 May 2019]. A review of the mental health challenges faced by academic staff, showing a lack of data but strong grounds for concern.

Levecque K, Anseel F, De Beuckelaer A, Van der Huyden J, Gisle L. 2017. Work organization and mental health problems in PhD students. *Research Policy* 46: 868–879. https://doi.org/10.1016/j.respol.2017.02.008. A study of mental health in Belgian PhD students showing one in two experiences psychological distress and one in three is at risk of a common psychiatric disorder, especially depression.

Sohn E. 2018. How to handle the dark days of depression. *Nature* 557: 267–269. https://doi.org/10.1038/d41586-018-05088-y. Stories and advice about mental health from researchers.

to experience and the ability to recognise what we don't know. It also requires an ability to set goals and to be thorough, methodical and rigorous. The demands can be extreme, ranging from remote fieldwork to challenging statistical analysis, and no single primatologist can be expert in all areas, making it essential to work together with other researchers.

1.3 THIS BOOK

I begin this book with several topics that are relevant throughout the scientific process: ethics (Chapter 2), research integrity (Chapter 3),

Box 1.4: Imposter syndrome (feeling like a fraud)

Imposter syndrome is the feeling that you don't deserve your success, that you're not as good as other people in the same position, that you're not good enough and that you might be exposed as a fraud, or disappoint your supporters. It is common in research, in which we receive critical feedback on work that is often very important to us.

Imposter syndrome can affect anyone, but is particularly prevalent in women, people of colour, and first-generation students. It may partly explain the under-representation of these groups in higher positions in science. Imposter syndrome affects those marginalised by society because it is difficult to believe in your own abilities when society assumes you are inferior and where there are few role models. Search for images of *professor* on the Internet, and you'll see what I mean (professors are overwhelmingly older white men). Imposter syndrome also affects those who feel that they are expected to do very well due to their background or circumstances.

Imposter syndrome skews your perspective on your own achievements and leads you to discount your own accomplishments, attributing them to luck, while you attribute other people's success to ability. If you score 98% on a test, imposter syndrome makes you focus on the 2% you didn't get right, rather than the extraordinarily high mark. It makes you ignore any positive comments on your work and focus on those that confirm your own feelings of inadequacy.

The harsh self-criticism associated with imposter syndrome reduces your confidence in your work and can be paralysing. It can cause you to delay seeking feedback, because you want your work to be perfect before you show it to anyone (**perfectionism**).

If this sounds like you, then these tips may help:

1. Realise that you're not alone. Many of the researchers you most admire suffer from imposter syndrome, too. Labelling and describing your experience helps.
2. Talk about it. This can be difficult but talking to other people helps a great deal. Online forums can help, too.
3. Recognise the difference between feeling and fact. Just because you feel incompetent doesn't mean you are incompetent. Take a step back and be fair to yourself. How would others see you? Are your expectations realistic?

Box 1 4: (cont.)

4. Remember that it's normal to feel daunted by a complex task. It doesn't mean you can't do it. Break it down into smaller, achievable sub-tasks (Box 1.3).

5. Don't compare yourself with other people (you will always find someone doing better than you).

6. Save positive feedback and read it again. (I keep it in a separate folder.)

7. Allow yourself to make mistakes. That's how we learn.

8. Remind yourself that negative comments on a piece of work aren't about you as a person, and they're not about all of your work (at least, they shouldn't be if the reviewer has done his or her job properly).

9. Learn to sort out useful constructive criticism from destructive criticism. Evaluate negative feedback carefully. Ask yourself whether all aspects of the criticism are true, and what you can do to address those aspects. If the criticism is personal (in other words, *ad hominem* attacks, directed at you, not your work) or nasty, consider the motivations of the person giving it.

10. Recognise that while we're constantly aware of our own failings, we see only what others choose to share with us. In other words, maybe everyone else is faking it, too. Success is the visible tip of an iceberg. You don't see what lies below the water. Perfection is impossible, and failure is not a catastrophe, even if it feels that way at the time.

11. Be generous with yourself, and with others. Researchers can be highly critical of one another and our working environment can be extremely competitive, but it doesn't need to be that way. Seek out collaborative and supportive colleagues.

12. Take care of your physical and mental health. Protect your life outside your research and take breaks. Take advantage of counselling if it's available to you.

This type of advice is easy to give, but hard to put into practice. Much of it puts the responsibility on the person with imposter syndrome. However, the negative ideas that we have about ourselves are internalisations of the power dynamics around us. Empower yourself by understanding the ways in which your struggles are not

---

Box 1.4: (cont.)

a reflection of your true worth, but a manifestation of the larger power dynamics in which we live and work.

If you don't have imposter syndrome, many of those around you have it. Take it seriously. Educate yourself about the power dynamics that underlie imposter syndrome. Look out for imposter syndrome in your friends and colleagues. We can all help by:

1. Being aware of our **social privilege**, which can make us think something is not a problem because it's not a problem to us personally.
2. Standing up for others who might be being bullied or facing extra obstacles.
3. Talking openly about imposter syndrome and supporting our colleagues.
4. Admitting when we don't know something and sharing failure as well as success.
5. Considering our own behaviour. If you habitually make fun of people, consider the effects this may have on those with imposter syndrome.
6. Being generous. Give positive feedback, as well as critical comments, on other people's work.
7. Being kind. If you must give someone's work a poor review, you don't have to do it in a harsh way; consider how it might affect them and encourage them to try again.

---

inclusive science (Chapter 4), understanding statistical evidence (Chapter 5) and communicating ideas in writing (Chapter 6). Having reviewed these essential themes, I introduce the primates (Chapter 7) and why we study them (Chapter 8).

The rest of the book works through the scientific process (Figure 1.1) as applied to primatology, from a general interest in a topic to reporting the study. First, I explore how we identify and develop a research question (Chapter 9). A thorough understanding of what we know and what we don't know, about a research area is essential to this process, so I explain how we search (Chapter 10) and read (Chapter 11) the scientific literature.

Understanding existing knowledge allows us to refine our research question and develop a set of hypotheses that propose possible answers to the question, derive testable predictions of exactly what we would

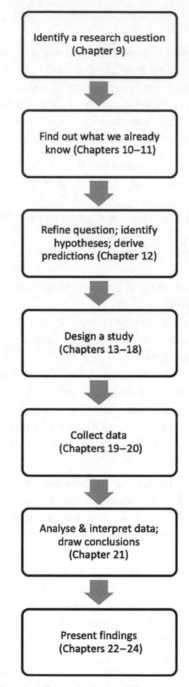

Figure 1.1 The scientific method, simplified.

expect to observe if a hypothesis is true and design a study to test those predictions (Chapter 12). We test predictions by observing natural variation or manipulating variables (Chapter 13) and making reliable and valid measurements (Chapter 14). Our study design dictates the statistical analysis we will use to test our predictions, and we need a detailed plan for statistical analysis (Chapter 15), using an appropriate sample size (Chapter 16). Finally, consideration of feasibility (Chapter 17) helps to refine our plans. Conveying all of these aspects of a study in a detailed proposal is essential for funding applications (Chapter 18).

Once we have a detailed study design (and funding), we can collect data (Chapter 19). Primatologists collect data in many settings, each with their own requirements, but the demands of fieldwork merit a separate chapter (Chapter 20). Data analysis and interpretation then allow us to test our hypotheses, interpret our results, compare them with what we already knew and draw conclusions about our original research question (Chapter 21).

When we have finished our study, we write it up as a scientific report (Chapter 22) and publish it (Chapter 23). Presentation at a scientific conference (orally or as a poster) is an excellent way to obtain feedback on our findings and inspiration for future work (Chapter 24). Chapter 25 provides conclusions and recommendations for primatologists.

I include few specific examples or case studies, for three reasons. First, readers can be distracted by the details of a specific example, rather than the general point I hope to convey. Second, there are plenty of examples in the literature. Third, I wanted to keep the book to a sensible length. Where I do include examples, they are often behavioural, due to my research background, but the generalities apply to all branches of primatology and to allied disciplines.

In most chapters I use *we* to describe what we do as a community of primatologists (including you and me), but for some chapters, I found that addressing the reader directly as *you*, or a mixture of the two, worked better.

Science is not as simple as Figure 1.1 and the progression of these chapters suggests. The stages of the research process feedback on one another in an iterative process. For example, we return to the literature regularly as we refine our plans. We revisit earlier decisions in the light of new insights, or when we encounter a practical problem. We might find ourselves at a dead end, when we can't obtain the data we require to test a prediction. We often abandon an idea and move back a step or restart the whole process. This is a normal part of science, even if it's not apparent from reports we read or presentations we attend. It happens to us all.

1.4 CHAPTER SUMMARY

In this chapter, we've seen that:

- Science seeks to explain observations of the natural world so that we can predict future observations.
- Scientists generate, then test, simplified models of the world.
- We correct our understanding based on what we find.
- We use replication to test the validity and generality of conclusions.
- Primatologists need a good understanding of theory, our study animals, and study design.
- We also need the ability to communicate in writing.
- We must be prepared to accept and act on critical feedback to refine and improve our work.
- Imposter syndrome is common among scientists.
- A well-designed study requires a great deal of work prior to data collection.
- The stages of the research process feedback on one another and it is normal to abandon ideas and start again.

In the next chapter, we look at the ethical issues we need to consider when designing and conducting a project.

1.5 FURTHER READING

Altman DG. 2012. Building a metaphor: Another brick in the wall? *British Medical Journal* 345: e8302. https://doi.org/10.1136/bmj.e8302. A comment on bricks and building metaphors for science.

Ayala, FJ. 2009. Darwin and the scientific method. *Proceedings of the National Academy of Sciences of the United States of America* 106: 10033–10039. https://doi.org/10.1073.pnas.0901404106. Describes the hypothetico–deductive scientific method, with a focus on Darwin's work.

Davies NB. 2010. Birds, butterflies and behavioural ecology. In: *Leaders in Animal Behavior: The Second Generation*. Drickamer LC, Dewsbury D (eds.). Cambridge: Cambridge University Press. pp. 143–164. An autobiographical essay in which Nick Davies (Professor of Behavioural Ecology at the University of Cambridge and Fellow of the Royal Society of London, UK) describes rushing into data collection at the beginning of his PhD, only to be brought to a standstill when his supervisor asked what his hypothesis was.

Hailman JP, Strier KB. 2006. *Planning, Proposing, and Presenting Science Effectively*. Cambridge: Cambridge University Press. Chapter 1 covers how science works.

Martin P, Bateson PPG. 2007. *Measuring Behaviour: An Introductory Guide*. 3rd edn. Cambridge: Cambridge University Press. Chapter 3 outlines the steps involved in the research process, focussing on studying animal behaviour.

Moreno E, Gutiérrez J-M. 2008. Ten simple rules for aspiring scientists in a low-income country. *PLoS Computational Biology* 4: e1000024. https://doi.org/10.1371/journal.pcbi.1000024. Proposes a set of rules for scientists to meet the challenges of building a scientific career in a low-income country.

Nettle D. 2018. *Hanging on to the Edges: Essays on Science, Society and the Academic Life*. Cambridge: Open Book Publishers. http://dx.doi.org/10.11647/OBP.0155. Essays on the highs and lows of life as an academic and a scientist, and on how to cope. Available for download free from the publisher.

Ruelas Inzunza E, Salazar-Rivera GI, Láinez M, Ruiz-Gómez MG, Domínguez-Eusebio CA, Cristóbal-Sánchez G, et al. 2017. Ten simple rules for successfully completing a graduate degree in Latin America. *PLoS Computational Biology* 13: e1005682. https://doi.org/10.1371/journal.pcbi.1005682. Advice for students interested in postgraduate programmes in Latin America, including funding.

Tinbergen N. 1963. On aims and methods of ethology. *Zeitschrift für Tierpsychologie* 20: 410–433. https://doi.org/10.1111/j.1439-0310.1963.tb01161.x. Outlines the four major questions we need to answer to understand a biological trait, two of which concern proximate mechanisms and two of which are ultimate in nature. The article focusses on behaviour, but the questions are just as useful for other traits.

Wright G. 2017. *Academia Obscura: The Hidden Silly Side of Higher Education*. For the lighter side of the academic experience. See also @AcademiaObscura on Twitter.

2

Ethics in Primatology

Ethics (or moral philosophy) is a branch of philosophy that addresses concepts of right and wrong. Ethics can describe personal beliefs or the rules and guidelines we use to establish whether conduct is right or wrong. It is our duty to consider the possible consequences of our work and mitigate any risks to the welfare and interests of our study animals, human participants, the environment, and the people we work with and alongside. We must also consider the effects of our research on our discipline and wider society. Reflecting on ethical issues and weighing the positive and negative impacts of a project are essential in making informed decisions when planning a project and throughout a study. This can include the decision not to conduct a study, or to terminate it earlier than planned.

In this chapter, I introduce approaches to ethics, cover legal requirements and permits, then address the ethics of working with primates in captivity and the wild, specimen collection, and working with human participants. I outline our ethical responsibilities to the natural environment, the people we work with and the people we work alongside. I highlight the importance of reflecting on our use of social media and the power of images, and end with our obligations to report and disseminate our findings.

2.1 APPROACHES TO ETHICS

Approaches to ethics in research include conformity to legal regulations, the bioethical principle of *do no harm*, justice, and benefi- cence. Scientists often employ cost–benefit or utilitarian analysis, focus- sing on the consequences of choosing one option over another. This consequential approach involves working out what the results of our research might lead to in terms of goods or harms to primates and others in the community.

Ethics are multilayered and complex. Our personal ethics reflect our background and experience. As primatologists, we are expected to follow ethical guidelines and codes of conduct. We also often work in communities who may have different ethical perspectives.

Guidelines cannot cover all eventualities, but instead teach us how to think. The most difficult ethical cases are where there are conflicts between different goods. For example, when the good of one species clashes with that of another in medical research, when the good of a primate species clashes with that of a community, or when the good of a species (in terms of conservation awareness) clashes with that of individual animals who experience the costs when we habituate wild animals for tourism and research.

2.2 LEGAL REQUIREMENTS AND PERMITS

We usually need permission to conduct research. This includes national, state, and institutional requirements. We must understand and comply with the statutory requirements and regulations of all countries in cross-national research. Local collaborators can help us to understand which permits we need.

Obtaining permissions can be very time-consuming and some researchers don't respect the requirement for research permissions. Don't copy them. Funding organisations may require copies of relevant permits before they transfer funds, and scientific journals require a statement that the research they publish complies with ethical guidelines and with the legal requirements of the host institution and the country or countries in which the research was conducted.

We need export and import permits to move biological samples of animals and plants across international borders. These include national paperwork from both countries and Convention on International Trade in Endangered Species of Wild Fauna and Flora (**CITES**) permits. We may also need to certify that samples are not infectious.

Some countries do not allow the export of biological samples. Host country collaborators can advise on this in cross-national research.

2.3 WORKING WITH CAPTIVE PRIMATES

Research on captive animals must follow the legal requirements of the country where the work is conducted, and all institutional guidelines.

Consider these requirements as the minimum and treat all animals as well as possible. Many countries and institutions require review and approval of projects by animal welfare, animal care, or ethics committees. Regardless of whether this is the case, we should follow relevant ethical guidelines published by national and international organisations such as the *International Primatological Society* and the *Association for the Study of Animal Behaviour/Animal Behavior Society*.

Recommendations for research on captive animals are usually based on the **3Rs** of Replacement, Reduction, and Refinement. **Replacement** means replacing animals with non-animals to achieve the same aims. **Reduction** involves reducing the number of animals used to the minimum required to test the hypotheses rigorously, by careful choice of study design. **Refinement** means designing a study carefully to minimise suffering and risk. This includes using the least invasive methods available, using positive reinforcement training rather than aversive stimulation, deprivation, or resource restriction to motivate animals to participate in a study, and minimising social disruption. Consult with experienced animal care staff who are familiar with the animals you want to study.

We harbour many pathogens that can be transmitted to other primates, including diarrhoea and respiratory diseases. To reduce the risk of transmission, wear gloves to handle animals, disinfect any apparatus that animals come into contact with after each testing session, make sure any food rewards are cleaned appropriately and handled with gloves and stay away if you are ill.

Animal care regulations also cover the procurement of animals, their transport, housing and care, and what happens to them once the project ends.

2.4 WORKING WITH WILD PRIMATES

Projects that involve capturing and handling wild primates, holding them in captivity, applying collars, marking them, collecting invasive biological samples, conducting surgical procedures (e.g. to implant recording devices), collecting biological specimens, and conducting field experiments are subject to the same regulations and guidelines as those that involve captive animals. In other words, they are subject to the 3Rs. We must carefully weigh the benefits to science against the costs to the animals and the ecosystem.

We usually study primates because we are interested in primates, so in most cases we can't easily replace them with non-animals. We can, however, replace animals in a study by using existing data and samples to achieve our goals. Reduction is more easily applicable; we should determine our sample size carefully such that it is the smallest neces- sary to test our hypotheses rigorously, particularly when we subject animals, their groups, population, or ecosystem to any sort of risk. Refinement involves employing the least disruptive research design required to achieve our aims, minimise risk to the animals, their groups, and the population. This includes only capturing animals when there is no less invasive way to obtain the information or biological samples we need. For example, where possible, we should use pheno- typic features to identify animals rather than capturing them to mark them and use non-invasive sampling rather than tissue sampling.

Endangered species may be subject to additional legislation and we should not place them at risk unless the study has a clear benefit for their conservation.

If we do need to capture wild animals, we should seek advice from researchers and veterinarians experienced in capturing primates under similar conditions. We must minimise direct contact with captured animals and use validated protocols to prevent disease transmission. We must minimise animal stress and suffering in all procedures and treat captured animals with care and respect. We should have a detailed plan of action or intervention in case an animal is injured and requires veterinary care or euthanasia. We should collect data on each capture we perform and use those data to improve protocols and contribute to collective efforts to improve capture methods for primates. Finally, we should maximise the scientific information that we obtain from cap- tured animals by collecting as many measurements and samples as possible without posing additional risk to the animal.

Animal care forms may not address non-invasive studies of pri- mates. Nevertheless, we must carefully consider the influence of the presence of human observers, habituation, and any provisioning on our study animals. For example, unhabituated animals may be stressed by the presence of observers. Habituation may be stressful to the animals and may increase their vulnerability to hunting, particularly when researchers are not present. Provisioning can provoke aggression, alter ranging patterns, and lead animals to depend on handouts. Our presence, habituation, provisioning, or presentation of foreign objects can also increase the risk of disease transmission from humans to animals. Thus, we should weigh the costs and benefits of our proposed study carefully. The 3Rs still apply. For example, replacement could involve studying previously habituated animals rather than subjecting additional animals to the risks of habituation. Reduction involves carefully considering and justifying our sample size. Refinement includes minimising stress during habituation, considering alternative approaches, and reducing the risk of disease transmission to our study animals.

To reduce the risk of disease transmission, we should:

- Ensure that all researchers, field assistants, and staff have up-to-date vaccinations against yellow fever, polio, tuberculosis, hepatitis, and other infectious diseases.
- Ensure that all researchers, field assistants, and staff take anti-parasite medication regularly and avoid exposing animals to human waste.
- Maintain high standards of hygiene and waste disposal at research camps.
- Respect quarantine periods after travel.
- Avoid close proximity to wild animals if we have symptoms like coughing, sneezing, fever, and diarrhoea.

For field experiments, we should also:

- Disinfect any apparatus that animals come into contact with after each testing session.
- Ensure that any food rewards are cleaned appropriately and handled with gloves.

Researchers at great ape research sites often wear face masks. In addition to minimising the risk of disease transmission (if appropriate masks are used correctly), this practice may help study animals to distinguish researchers from hunters and helps to communicate an appropriate message to tourists and the public.

2.5 COLLECTING BIOLOGICAL SPECIMENS

Animal remains have enormous scientific potential for systematics, biogeography, genomics, morphology, parasitology, stable isotope ecology, and other research areas. If you find remains, first consider the risk of pathogen transfer. If it is safe to do so, collect, document, and preserve the remains carefully, then deposit them in a referenced collection, with as much information as possible, ideally in the same country.

We have an ethical obligation to maximise the information gained from animals that are killed accidentally or euthanised. Traditional field biology also included killing animals to collect them as biological specimens. This practice is now heavily debated and subject to strict permitting regulations. Again, the 3Rs apply. Possibilities for replacement include waiting to find an animal that has died naturally and using existing specimens. Reduction involves reducing numbers to the minimum required for good science, which may be zero. Possibilities for refinement include non-lethal alternatives to collection (e.g. photographs, tissue samples, recordings, non-lethal and non-invasive sampling). Obviously, we should never collect specimens if doing so threatens the survival of a population or a species.

Never purchase animals for sample collection, because it promotes demand for wild animals. Simply expressing interest in animals can promote capture of wild animals because people do not always understand our motives.

It can be tempting to maximise the information gained from animals culled by authorities because they are perceived as pests. However, this requires careful thought; our scientific research may be used to justify further culling of animals.

2.6 WORKING WITH HUMAN PARTICIPANTS

Research involving human participants is subject to ethical oversight and approval from our institution and national bodies. This is separate from, and different from ethical review for work with animals and includes careful consideration of how data are collected, handled, and stored. Ethics review involves protecting research participants, anticipating harms, avoiding undue intrusion, negotiating informed consent, and respecting participants' rights to confidentiality and anonymity. This process requires careful reflection and may be less familiar to primatologists than animal care and use forms. It covers all work with humans, including formal and informal interviews, observations, and sampling.

Research involving vulnerable individuals and children requires particularly careful ethical consideration. You may need background checks and evidence that you have no criminal record. It can take time to obtain these.

2.7 WORKING IN THE NATURAL ENVIRONMENT

Field studies can have negative consequences for the local environment, including disturbance to local wildlife, the introduction of human waste and refuse, and the effects of trails and trail traffic on vegetation and soil. We should disturb habitats as little as possible. We should minimise injury to vegetation, for example when labelling trees. We should minimise vegetation cutting, especially of mature trees, when making trails and walk through mud, not around it, to reduce soil erosion on trails. Well-used field stations may need to lay paths on major trails. We should use biodegradable materials to mark trails or trees if our research is short-term, and collect flagging tape, and so forth, when we no longer need it.

Establishing a new field site can direct attention to an area and the animals that inhabit it and open it up for subsequent access. We must consider what will happen when we are not present. Are we putting the animals, plants, or environment in danger?

2.8 WORKING ALONGSIDE OTHER PEOPLE

Studying primates always involves working with and alongside other people. These people may include our hosts, the staff, and other researchers working at the same research site, employees, and volunteer research assistants. Our working relationships may be with friends or with people we wouldn't otherwise choose to spend time with. In all cases, working relationships require respect and sensitivity, openness to other people's opinions, understanding of their priorities, and good communication. We must also acknowledge economic and other inequalities and their effects on the opportunities available to people.

Conflicts can arise in working relationships due to different priorities and misunderstandings. We should question our own preconceptions and understand that people may make assumptions about us based on previous experience or stereotypes. We can't always avoid conflict, but we can anticipate and mitigate it by being positive and respectful and communicating effectively to develop trust and rapport.

If we employ people, this makes us a manager, so we should learn management skills. These include providing clear expectations, listening to employees, communicating effectively, and offering opportunities for personal and career development. We should draw up a contract and train and mentor the people we employ appropriately. We should explain the project to them and answer any questions they have. We must take responsibility for the health and safety of all employees and volunteers at work, including risks of injury, accident, and exposure to chemical hazards and to pathogens.

If we take on unpaid, or low-paid, research assistants, we must remember that they are working for little or nothing or even paying to help us with our work and treat them accordingly. Relying on unpaid or paying interns propagates inequality in our field, as many people can't afford such opportunities. We should include appropriate salaries and compensation in funding applications.

If we hire assistants from the area around our study site, or students from nearby universities, we should seek advice from colleagues experienced in the local area about how to select and manage people and appropriate pay scales. Employing people can lead to jealousy and resentment in the local community. We must consider local expectations and norms, whether what we ask of our assistants conflicts with these (e.g. expecting men and women to work together in the forest at night), and who bears the costs of any conflicts.

If the local language at our research site is different from our own, we should learn at least a few pleasantries, to show respect and break down barriers. Learning to communicate well in the language will improve working relationships greatly and help to avoid misunderstandings. If we work abroad, we must respect the fact we are guests in another country. We should not assume things will, or should, work the same way that they do in our own country.

We should seek opportunities to share our knowledge with others at our research site, including students and staff. If we are from a high-income country and we work in a low- or middle-income country, we should consider the interests of local students and researchers and appreciate how disparities in access to resources influence equity in collaboration. Some countries require visiting researchers to include and train nationals in their projects. Where appropriate, we should advise and mentor local scientists to help them gain entry to a field where foreigners are gatekeepers to resources and opportunities. We must not patronise the people we work with, or assume we know better or more than they do.

2.9 COLLABORATING EFFECTIVELY

Collaboration means working together to achieve a common goal. Thus, anyone who contributes to a research project is a collaborator. Researchers often prioritise intellectual contributions over practical input to a study, but both are essential to a project. Technical staff and assistants often provide new ways of thinking about questions, study design, and interpretation that we have not otherwise considered. Payment for services and power relationships also influence who is deemed to be a collaborator. Consider these issues carefully when describing collaborations.

Collaborations occur between students and supervisors, postdoctoral scholars and mentors, hosts and visiting researchers, and researchers and non-academic partners. They may involve large teams or combine multiple groups. Large collaborative teams need careful management. In some cases, we already know our collaborators, which makes suggesting a project relatively simple. In others, we might approach someone we have never met to suggest a project.

Collaborations allow us to answer questions we could not otherwise address and to discover entirely new questions. They bring us fresh ideas, intellectual stimulation, and access to facilities and resources. Collaboration is also powerful when we discover that another group is working on similar questions to us because we make more progress by working together than we do by competing. However, when collaboration goes wrong, it can lead to nasty disputes, wreck projects, tarnish reputations, and end friendships.

Collaborations are often cross-disciplinary and cross-national. This diversity of backgrounds makes it essential to understand the perspectives of the people you're working with. Successful collaboration requires teamwork and trust, mutual respect, an open mind, willingness to accept other people's views and experience, acknowledgement of economic and other inequalities, respect for differences in timescales people work at and differences in objectives, and recognition that collaborators bring different, but essential, resources to the project.

Discuss and agree on expectations at the beginning of a project, including the role and contribution of each partner, and what each is responsible for. A written agreement, or **Memorandum of Understanding** is very useful. This should include who does what and when; who pays for what; who owns which data; who will be an author of any presentations and reports, and the order of authors; what happens if a project partner moves on; and what happens if you want to follow-up on the project. This might seem excessive, but it will avoid awkward situations when you realise that your understanding of a verbal (or assumed) agreement is very different to that of your collaborator.

Communicate throughout the project and meet face-to-face, if possible. Keep an agreed-upon record of discussions and review it together regularly, updating your agreement as the project progresses. Don't assume things will work out or that you share the same view of an issue if you haven't discussed it. Ownership of research ideas can be a particular sticking point, as can what does, and does not, merit authorship.

2.10 LIVING IN OR NEAR COMMUNITIES

If a project involves fieldwork, establishing good relations with the people around the study area is essential to success. Consult with the local communities and explain your project to them before you begin your research. This goes well beyond obtaining formal research permissions. Negative interactions with local communities can jeopardise future research efforts.

Learn about and respect the history and culture of the region you work in and the peoples who live there. Be sensitive to cultural differences and to how you are perceived. In other words, reflect on your own individual and cultural values, beliefs, preferences and expectations, and seek to understand those of the people around you. Be humble.

Purchase supplies from the local community where possible, but don't buy up the entire stock of something other people may need, too. Discuss your needs with local suppliers.

Think carefully about the possible effects of your project and any findings on local human communities. Remember that the needs of local people may conflict with the needs of study animals.

At some field sites, we risk transmitting pathogens to people who have not been exposed to them. Ensure that you are vaccinated and follow protocols to reduce disease transmission.

2.11 USING SOCIAL MEDIA

Social media plays a large role in how we communicate. Think carefully before sharing images on social media or in presentations. Work in captive environments is often sensitive and your host institution may not permit unauthorised sharing of material. Ask for consent before sharing images of people, and take opportunities to challenge, rather than reinforce, negative stereotypes. Avoid images that portray people as helpless or passive. Use captions to provide details and context and remember that images travel further than their captions, so your explanation may not reach people who view the image.

Everything we do as a primatologist influences primate welfare. In other words, we are ambassadors for primates.[1] Viewing images of people in close proximity to primates makes people less likely to perceive primates as endangered and increases the likelihood that people consider primates as appealing pets. Don't post or publish such images and don't use them in presentations. This includes photographs from sanctuaries. Photographs are often viewed in isolation and selfies with wild animals (even if you're helping them) unwittingly send the message that it's okay to approach and touch wild animals, or even to keep them in captivity.

2.12 REPORTING AND DISSEMINATING OUTCOMES

Consider any potential for harm to your study subjects, research participants, and local people when making data available. Include a note on ethics when reporting on your project.

Funders expect final reports on your project. You should also communicate your research findings in a suitable format and language to your host institutions and facilities, local governments, and other interested groups or communities, as appropriate.

Think carefully about the ways in which you disseminate your findings in the popular press. Avoid sensationalist language that may have negative consequences for primates or the credibility of scientific research.

2.13 CHAPTER SUMMARY

In this chapter, we've seen that:

- We have an ethical imperative to avoid harm to our study animals, human participants, the environment, the people we work with and alongside, our discipline, and wider society.
- Ethical guidelines cannot cover all eventualities, but instead teach us how to think about ethical issues.
- Our work is subject to legal requirements, permissions, and institutional approval.

---

[1] With thanks to Hannah Buchanan-Smith for the concept.

- Work with captive primates is governed by animal use committees and the 3Rs of replacement, reduction, and refinement.
- The 3Rs also apply to work with wild primates, including observation-only studies.
- We should employ the least disruptive research design required to achieve our aims, minimise risk to the animals, their groups, and the population.
- We must maximise the information we obtain from captured animals.
- We must consider and reduce the risk of transmitting disease to animals and local human populations.
- We must consider alternatives to biological specimen collection.
- All research involving human participants requires appropriate ethical review, informed consent, and respect for rights to confidentiality and anonymity.
- We must minimise disturbance to the natural environment in field projects.
- Science involves working with other people and these working relationships require respect, sensitivity, and openness to other people's opinions.
- Formal agreements are very useful when managing collaborations.
- We must consider the implications of our project for people living in local communities.
- We are responsible for the reputation of our discipline. Our actions should not jeopardise future research efforts.
- We must reflect carefully on the images we share.

In the next chapter, we look at scientific integrity, a topic closely related to ethics.

2.14 FURTHER READING

Association of Social Anthropologists of the UK and Commonwealth. *Ethical Guidelines for the Conduct of Anthropological Research*. www.theasa.org/ethics/guidelines.shtml [Accessed 2 May 2019]. Includes relations with and responsibilities to human informants, colleagues, and wider society.

Association for the Study of Animal Behaviour/Animal Behavior Society. 2018. Guidelines for the treatment of animals in behavioural research and teaching. *Animal Behaviour* 135: I–X. https://doi.org/10.1016/j.anbehav.2017.10.001. Covers work with captive animals and fieldwork. Updated in each January edition of *Animal Behaviour*.

Brando S, Buchanan-Smith H. 2018. The 24/7 approach to promoting optimal welfare for captive wild animals. *Behavioural Processes* 156: 83–95. https://doi .org/10.1016/j.beproc.2017.09.010. A holistic approach to animal welfare, with a focus on zoos, wildlife centres and sanctuaries. See also 247animal-welfare.eu.

Code of Ethics of the American Anthropological Association. https://s3 .amazonaws.com/rdcms-aaa/files/production/public/FileDownloads/pdfs/issues/ policy-advocacy/upload/ethicscode.pdf [Accessed 2 May 2019]. Includes our responsibilities to people and animals that we study, scholarship and science, and the public.

Cunningham EP, Unwin S, Setchell JM. 2015. Darting primates in the field: a review of reporting trends and a survey of practices and their effect on the primates involved. *International Journal of Primatology* 36: 894–915. https://doi .org/10.1007/s10764-015-9862-0. Calls for greater sharing of information about primate capture and immobilisation to improve methods.

Gadlin H, Jessar K. *Preempting Discord: Prenuptial Agreements for Scientists.* US Office of Research Integrity. https://ori.hhs.gov/preempting-discord-prenuptial-agreements-scientists [Accessed 2 May 2019]. A blog post on misunderstanding in collaborations and key issues to address in a collaboration agreement.

Geissler PW, Okwaro F. 2014. Discuss inequality. *Nature* 513: 303. https://doi.org/ 10.1038/513303a. On collaboration under conditions of inequality and the need to confront economic differences. Focusses on health, but applicable more broadly.

Gruen L, Fultz A, Pruetz J. 2013. Ethical issues in African great ape field studies. *Institute for Laboratory Animal Research Journal* 54: 24–32, https://doi.org/10 .1093/ilar/ilt016. Focusses on the ethics of studying African great apes, but the issues are relevant to all primates. Includes decision trees for habituation and intervention that can be applied to all primates.

Hubrecht RC, Kirkwood J. (eds.). 2010. *The UFAW Handbook on the Care and Management of Laboratory and Other Research Animals.* 8th edn. Oxford: Wiley-Blackwell. Key reference for the ethical use of animals in research.

Kilkenny C, Browne WJ, Cuthill IC, Emerson M, Altman DG. 2010. Improving bioscience research reporting: The ARRIVE guidelines for reporting animal research. *PLoS Biology* 8(6): e1000412. https://doi.org/10.1371/journal.pbio .1000412. Introduces a checklist of the minimum information that all scientific publications reporting research using animals should include.

MacClancy J, Fuentes A. (eds.). 2013. *Ethics in the Field: Contemporary Challenges.* New York: Berghahn. Includes chapters on the unintended consequences of primatological fieldwork (McLennan & Hill), the effects of observational field studies on primates (Strier), the various ethical perspectives we negotiate in field primatology (Kutsukake), key concerns in fieldwork ethics (MacKinnon & Riley) and the need to study species other than great apes (Nekaris & Nijman).

Nordling L. 2018. Europe's biggest research fund cracks down on 'ethics dumping'. *Nature* 559: 17–18. https://doi.org/10.1038/d41586-018-05616-w. Reports on a new code of conduct for European Union-funded research to tackle the practice of conducting ethically dubious research in foreign countries.

Radi-Aid. *How to Communicate the World. A Social Media Guide for Volunteers and Travellers.* www.rustyradiator.com/social-media-guide/ [Accessed 2 May 2019]. Includes a checklist to consult before posting on social media. Also see barbiesavior.com for a parody of volunteer images.

Riley EP, Bezanson M. 2018. Ethics of primate fieldwork: Toward an ethically engaged primatology. *Annual Review of Anthropology* 47: 493–512. https://doi

.org/10.1146/annurev-anthro-102317-045913. Detailed consideration of ethical practice in field primatology, and how to think critically about the consequences of our research including the notion that conducting primate fieldwork is a privilege, not a right.

Riley EP, Mackinnon KC. (eds.). 2010. Special section on ethical issues in field primatology. *American Journal of Primatology* 72(9): 749–840. A collection of commentaries on and a review of ethical issues faced by field primatologists.

Riley EP, Mackinnon KC, Fernandez-Duque E, Setchell JM, Garber PA. 2014. *Code of Best Practices for Field Primatology*. Highlights a set of ethical issues that should be considered in conducting field research and a set of practices that could be employed when confronting those ethical issues. Includes links to additional information and ethical guidance. www.internationalprimatologicalsociety.org/docs/code%20of_best_prac tices%20oct%202014.pdf [Accessed 2 May 2019].

Ross SR, Vreeman VM, Lonsdorf EV. 2011. Specific image characteristics influence attitudes about chimpanzee conservation and use as pets. *PLOS ONE* 6: e22050. https://doi.org/10.1371/journal.pone.0022050. Provides evidence that the presence of a human in a photograph with an ape increases the likelihood that people consider wild populations as stable and healthy and consider chimpanzees appealing as a pet.

Scully EJ, Basnet S, Wrangham RW, Muller MN, Otali E, Hyeroba D, et al. (2018). Lethal respiratory disease associated with human rhinovirus C in wild chimpanzees, Uganda, 2018. *Emerging Infectious Diseases* 24: 267–274. https://doi .org/10.3201/eid2402.170778. Evidence that a virus that causes the common cold in humans can kill chimpanzees.

Smith AJ, Clutton RE, Lilley E, Hansen KEA, Brattelid T. 2017. PREPARE: Guidelines for planning animal research and testing. *Laboratory Animals* 52: 135–141. https://doi.org/10.1177/0023677217724823. Planning guidelines for animal research. See also: https://norecopa.no/prepare.

World Animal Protection. *A Close Up on Cruelty: The Harmful Impact of Wildlife Selfies in the Amazon*. www.worldanimalprotection.us.org/sites/default/files/us_files/ amazonselfiesreport_us.pdf [Accessed 2 May 2019]. On the link between wildlife selfies and animal cruelty. Selfies with our study subjects can lead viewers to think that it's okay to approach or even be in contact with wild animals.

World Conference on Research Integrity. 2013. *Montreal Statement on Research Integrity in Cross-Boundary Research Collaborations*. www.researchintegrity.org/ Statements/Montreal%20Statement%20English.pdf [Accessed 9 January 2019]. Provides guidance on the conduct of research collaborations between different institutions, disciplines, sectors, and countries.

3

Keeping Science Healthy: Research Integrity

Good scientists are sceptical and seek evidence that contradicts a hypothesis (Chapter 1). Research integrity means conducting science in such a way that others can be confident in the methods we used and trust the findings we report. In addition to our responsibility to understand and comply with the ethical and legal obligations associated with our research (Chapter 2), research integrity involves scrupulous honesty and the highest standards of rigour. However, a combination of our own biases, distorted career incentives, poor understanding of study design and misuse of statistical analysis lead to practices that damage science (**questionable research practices**). Such practices undermine the validity of studies and increase the chance of erroneous results, leading to a literature based on false-positive conclusions and studies that can't be replicated. The inability to replicate the findings of published studies has been popularised as the **replication crisis**, particularly in medicine and psychology.

In this chapter, I first define research misconduct and its consequences. I then review responsible practices and how to avoid questionable research practices. We'll revisit these issues throughout the book.

3.1 RESEARCH MISCONDUCT

Research misconduct, or **research fraud**, undermines the research integrity of the individual, group, institution, and field concerned. It includes fabrication, falsification, and plagiarism. It does not include honest error.

Fabrication means making up data.

Falsification means manipulating or selecting data so that the research is not reported accurately. It includes manipulating images and changing or omitting data.

Plagiarism means appropriating another person's work and presenting it as though it is your own without full acknowledgement. It includes copying ideas, phrases, and sentences from published and unpublished material in articles, abstracts, conference presentations and grant proposals, and on the Internet. Plagiarism includes altering a few words or phrases in a sentence or changing the order of phrases. It also includes recycling text written by a collaborator.

Plagiarism can be a deliberate attempt to steal ideas, which is clearly unethical. Never use ideas, data, or methods you learn about when reviewing a grant application or a manuscript without permission from the authors. Never use ideas that someone has discussed with you without their permission.

Plagiarism can also stem from a lack of confidence in your own writing, particularly if English is not your first language. This is easily picked up by plagiarism detection software, and often appears as a patchwork of sentences from various sources or changes in the quality of the English language between sections of a report. Citing the original author does not solve the problem. Although the reasons for this type of plagiarism are understandable but it can get you into a lot of trouble. To avoid plagiarism, always express ideas in your own words and always credit the original author. Practice rephrasing. If you quote material verbatim, use quotation marks, and cite the original author.

Research misconduct has serious consequences. Researchers found to have committed research misconduct may be barred from applying to funding agencies, banned from publishing in journals, or dismissed from an institution. Students may fail their degree and universities can revoke degrees if they later discover misconduct.

If editors of scientific journals suspect research misconduct, they investigate. If they find evidence of research misconduct in a submitted manuscript, they reject it. If they determine that a problem with a submitted manuscript is a result of honest error, they may invite the author to revise the manuscript.

If editors find evidence of fraud in a published article, they retract it to correct the literature. **Retraction** involves the publication of a notice that is permanently linked to the original article, stating that it should not have been published and explaining why. It's not possible to replace a published article with a new version because earlier versions may persist. We use retraction to correct the scientific record, not to punish the author.

3.2 REVIEW THE LITERATURE FAIRLY, ACCURATELY, AND APPROPRIATELY

Literature reviews summarise and synthesise existing work on a topic. They are a crucial part of funding proposals and research articles. They should be based on a detailed literature search. They should acknowledge and credit relevant work by citing other researchers' work fully, fairly, and accurately. It is bad practice to review the literature selectively to make a study appear more novel than it is. It is also bad practice to cite only work from your own group, or that of friends, or to include irrelevant citations to boost someone's citation counts.

3.3 USE BLIND PROTOCOLS

When we expect a particular result, our beliefs can influence the way we collect data (the **experimenter effect**). In **blind protocols**, the data collector does not know which subjects are assigned to which treatments. Ideally, the data collector should also be blind to the study aim. In some cases, it's not possible to collect data blind. For example, it's often not possible to be blind to the sex of a primate, and our expectations of how animals of different sexes behave can influence our observations. Ensuring that observers agree on scoring (high **interobserver reliability**) reduces bias, although observers may share the same bias. We should discuss the potential for bias openly in reports.

3.4 CONCENTRATE ON RIGOUR, NOT STATISTICAL SIGNIFICANCE

The outcome of a study often influences whether researchers publish it (the **file-drawer effect** or **publication bias**). Results selected for publication may be those that reject a null hypothesis (more on this in Chapter 5), are consistent with a favoured hypothesis, or are surprising. As a result, the literature is biased towards studies that report statistically significant findings in favour of the proposed hypothesis (often termed **positive results**), while non-significant (or negative) findings are less likely to be published.

We can address this problem by concentrating on the rigour of a study, rather than the outcome. We should report studies regardless of the outcome, as long as they are designed appropriately to address the aims. Similarly, we should not base peer-review recommendations on the outcome of a study, but on the rigour with which it was conducted.

3.5 REPORT STUDIES HONESTLY AND OPENLY

Research that generates hypotheses is termed **exploratory**. Research that tests hypotheses is termed **confirmatory**. This distinction has important implications for how we assess research findings. Presenting exploratory research as confirmatory gives the reader false confidence in the findings and decreases the likelihood that they will be replicated. However, our cognitive biases, misunderstandings of the scientific process and career incentives make it easy to confuse the two. For example, if our predictions are vague, we can test them in a variety of ways, then select results that support our hypothesis, convincing ourselves after the event (*post hoc*) that these are the most appropriate tests to report. In other words, we confuse **postdiction** with prediction.

Presenting an analysis that is informed by the results (a *post hoc* **analysis**) as though it were based on theoretical deduction (an *a priori* **hypothesis**) is known as hypothesising after the results are known (**HARKing**), *post hoc* **theorising** or *post hoc* **storytelling**. The related practice of running multiple exploratory analyses of a dataset, identifying possible relationships, testing those relationships statistically using the same dataset and presenting the results as though the analyses were pre-planned is termed **data dredging**, **data mining**, **data fishing**, **data snooping**, or **double dipping**. This practice dramatically inflates the likelihood of inferring a pattern that does not exist. To avoid these problems, we should limit confirmatory analyses to pre-planned comparisons to test specific predictions, and state clearly where we conduct *post hoc* exploratory analyses.

The many decisions we make about data collection, analysis, and reporting during a study influence the study outcome and conclusions. Exploiting these **researcher degrees of freedom** to obtain a desired outcome – usually a statistically significant test of a null hypothesis – then reporting only that outcome, neglecting to report all the other analyses we conducted, is termed **selective reporting**, *p*-hacking, **cherry-picking**, **flexible analysis**, or **significance-chasing**. This is a serious problem, because it biases or distorts the scientific record. To avoid this problem, we should be honest, rigorous and transparent throughout the scientific process, including how we present our aims, report the methods we use, collect and analyse data, and interpret and communicate our findings. In other words, we should not mislead our audience.

Conducting multiple tests of the null hypothesis during data collection and stopping when the results are statistically significant (**data-peeking** or **optional stopping**) also distorts our results. Testing as we go along might also lead us to abandon data collection prematurely, because we don't find significant results early in a study. The solution to this is to select an appropriate sample size and leave data analysis until we have finished data collection. Alternatively, we can use statistical analyses that account for data-peeking.

The **Open Science** movement aims to remove barriers to sharing knowledge at all stages of the research process, promote transparency, and facilitate collaboration. It includes **preregistration** of research plans, reporting the entire workflow, and sharing data analysis code, making articles freely and permanently available online for anyone to access (**open access** publishing) and making data available for anyone to access and use (**open data**).

3.6 ADMIT MISTAKES

Everyone makes mistakes. The difference between misconduct and honest error lies in the researcher's intention. Misconduct involves deliberate deception, whereas honest error does not.

If we discover a mistake in a manuscript during the review and editorial process, before it is published, we should contact the editor and arrange to update the files. Once an article is published, we can't change the contents, because it is part of the scientific record. Instead, for minor errors, we publish a correction notice (a **corrigendum**), explaining the error, and correcting it. If the publisher introduced the error, the correction is termed an **erratum**, although publishers differ in the terms they use. Where an error means that findings of a study are unreliable, we **retract** an article. The original article will usually remain available, but it will be linked to a retraction notice, explaining the reason for retraction.

Publishing a correction or retracting an article due to honest error can be a daunting prospect, but corrections and retractions are essential to correct the scientific record. Publishing them where necessary reflects your scientific integrity. Good scientists will support you.

3.7 USE RESEARCH FUNDS APPROPRIATELY

Research funds are usually awarded for specific purposes. We should use them economically and only for the stated purposes. Some awards have particular conditions, which we must observe. Some funders allow transfer of funds between budget headings, but we must ask permission from the funder before doing this.

3.8 RESPECT THE PEER-REVIEW PROCESS

Peer review is the process by which a researcher's work is subjected to scrutiny and evaluation by experts in the same field (Box 3.1). It is a crucial component of science. We review one another's grant applications and manuscripts. Abuse of the peer-review process for personal gain includes copying ideas from a funding proposal or unpublished work, engineering citations of one's own work, providing a hypercritical review in an attempt to stop the funding or publication of a perceived competitor's research or deliberately delaying publication of someone else's work by stalling the review process.

3.9 ASSIGN AUTHORSHIP CREDIT APPROPRIATELY

We should recognise all contributions to a research project with authorship and acknowledgement, as appropriate (Box 3.2). Contributions to a project include: conceiving the original idea; designing the study; maintaining a field site, captive facility, collection or database; collecting data; analysing data; interpreting the results; and writing the manuscript. Thank contributors who do not meet the criteria for authorship in the **acknowledgements** sections of reports. For example, we should acknowledge people who provided advice, or critical reading of the manuscript.

The use of publications as a measure of productivity and currency in career advancement results in intense pressure to publish, making authorship contentious. Many researchers regard intellectual contributions as more important than practical contributions to a project when determining authorship. However, both are crucial to the success of a project.

Box 3.1: Peer review

Peer review is the independent evaluation of work, including funding applications and research reports, by our peers to assess its validity and quality. It is much criticised and is certainly not perfect. However, it is the only widely accepted method for validating research and serves to improve the final result.

The review process can be single-blind, double-blind, or open.

In **single-blind review**, the identities of the authors are known to the reviewers, but the identity of the reviewers is not known to the authors, unless they choose to sign their review. This is designed to allow reviewers to provide critical and constructive comments on a manuscript in the absence of personal consequences. However, single-blind review may allow biases based on the author's identity rather than on the contents of the manuscript to influence the review process. These biases may include relationship to the reviewer, gender, seniority, nationality, reputation, and affiliation.

In **double-blind review**, the reviewers do not know the identities of the authors, in addition to the authors not knowing the identities of the reviewers. Reviewers may be able to guess the identity of the authors, but these guesses can be inaccurate, and most manuscripts have multiple authors, meaning that the reviewer cannot identify the full team, or the position of the authors in the list. Employing double-blind review also serves to remind reviewers of the need to provide an impartial review.

In **open review**, reviews are published alongside an article. This has the benefit of acknowledging the important role of reviewers and encouraging in-depth reviewing. However, open review does not address the possible influence of the personal consequences of a flattering or critical review on the review process.

FURTHER READING

Polka J, Kiley R, Konforti B, Sterna B, Vale RD. 2018. Publish peer reviews. *Nature* 560: 545–547. https://doi.org/10.1038/d41586-018-06032-w. A call for open review of manuscripts, with a useful history of peer review.
Setchell JM. 2015. Double-blind peer review and the advantages of sharing data. *International Journal of Primatology* 36: 891. https://doi.org/10.1007/s10764-015-9860-2. Editorial announcing that the *International Journal of Primatology* has implemented double-blind peer review.

Box 3.2: Authorship

Authorship confers credit for an individual's contribution to published work. It also entails responsibility and accountability for the contents of that work.

Who Qualifies as an Author?
There are no simple rules governing authorship. However, the United States National Academy of Sciences recommends the following best practice, generalised from recommendations of the International Committee of Medical Journal Editors.

> Each author is expected to have made substantial contributions to the conception or design of the work; or the acquisition, analysis, or interpretation of data; or the creation of new software used in the work; or have drafted the work or substantively revised it; AND to have approved the submitted version (and any substantially modified version that involves the author's contribution to the study); AND to have agreed both to be personally accountable for the author's own contributions and to ensure that questions related to the accuracy or integrity of any part of the work, even ones in which the author was not personally involved, are appropriately investigated, resolved, and the resolution documented in the literature.

Participation in writing the grant that funded the research does not necessarily justify authorship, nor does heading the lab or site at which the work was conducted, although some institutions require this.

Authorship concerns include: guest or honorary authorship, where authors did not contribute to the research; gift authorship, where the contribution merits acknowledgement rather than authorship; and ghost authors, where those who do meet the criteria are excluded.

Order of Authors
There are various conventions for the order of authors on a manuscript:

- If the work is part of a student's thesis, that student is usually first author. If the student chooses not to submit the work for publication, or does not do so by an agreed date, the supervisor may be first author

Box 3.2: (cont.)

- The first author contributed most of the work
- Where the author list is very long, authors may be listed in alphabetical or reverse alphabetical order
- If the article is one of a series resulting from a collaborative project, authors may rotate the position of first author
- The last author is often assumed to be the driving force behind the research and is often the principal investigator, but this is not always the case
- *Joint* or *Co*-authors reflect an equal contribution to the manuscript. This is usually the first two authors but can be any combination of authors.

The Corresponding Author

The corresponding author takes primary responsibility for communication with the journal during the manuscript submission, peer review, and publication process. After publication, the corresponding author answers queries about the article and share data. He or she ensures that all listed authors have approved the manuscript prior to submission and that all authors receive correspondence with the editor about the manuscript, including full reviews. He or she is responsible for ensuring that all data are archived appropriately.

Many researchers and institutions associate the corresponding author with leadership on a project.

Author Contribution Statements

Author contribution statements explain how each author contributed to a report. Some journals require these statements. If a journal doesn't require this, you can include contributor information in the acknowledgements section.

Negotiating Authorship

Discussion of authorship is a dynamic process. The position of first author is most hotly contested, because it can be very important in career progression. For example, many systems require PhD students to publish at least one paper as first author, and institutions and funders use first author publications to evaluate researchers. Disputes usually arise because a person feels that their contribution has been undervalued or because they feel that they need a publication for their career.

Box 3.2: (cont.)

Begin by agreeing on responsibilities, roles, and expectations for authorship at the beginning of a collaboration. You may plan several manuscripts. Update the agreement as the project progresses. For example, you might add an element to the project, recruit a new collaborator, or increase or decrease the responsibility of a project participant. Changes to your publication plans include new manuscripts, including additional authors, removing authors, or rearranging the order of authors. Frequent and open communication minimises the potential for misunderstanding, disagreement, and resentment. A difficult conversation is better than later recriminations. In cases of serious disagreement, you may need to seek advice from your institution or journal editors.

Authorship discussions are particularly difficult when there are power differentials between project participants, for example between students and supervisors, or researchers and principal investigators. Disputes include disagreements over the importance of a project participant's contribution to a project, or who conceived an idea.

Authorship disputes also occur where two researchers, or teams, work on closely related topics or on the same materials or study subjects. Again, open and honest communication about plans is crucial to maintain working relationships.

ORCID

Open Researcher and Contributor ID **(ORCID)** provides persistent author identifiers that uniquely identify authors. These IDs avoid name confusion and provide a clear record of an individual's contributions as an author.

FURTHER READING

Albert T. 2003. How to handle authorship disputes: A guide for new researchers. *The COPE Report.* https://publicationethics.org/files/2003pdf12_0.pdf [Accessed 2 May 2019].

Brand A, Allen L, Altman M, Hlava M, Scott J. 2015. Beyond authorship: Attribution, contribution, collaboration, and credit. *Learned Publishing* 28: 151–155. https://doi.org/10.1087/20150211. Describes a contributor role taxonomy identifying specific contributions to published research.

Box 3.2: (cont.)

Clymo RS. 2014. *Reporting Research: A Biologist's Guide to Articles, Talks, and Posters*. Cambridge: Cambridge University Press. Chapter 4 covers authorship.

Council of Science Editors. *Authorship and Authorship Responsibilities*. www .councilscienceeditors.org/resource-library/editorial-policies/white-paper-on-publication-ethics/2-2-authorship-and-authorship-responsibil ities/ [Accessed 9 January 2019]. Principles to guide authorship-related decisions, policies, practices, and responsibilities.

Gaffey A. 2015. Determining and negotiating authorship. *Psychological Science Agenda*. www.apa.org/science/about/psa/2015/06/determining-authorship .aspx [Accessed 9 January 2019]. Includes examples of authorship contracts.

International Committee of Medical Journal Editors. *Defining the Role of Authors and Contributors*. www.icmje.org/recommendations/browse/ roles-and-responsibilities/defining-the-role-of-authors-and-contributors .html [Accessed 9 January 2019]. Defines criteria for authorship.

McNutt MK, Bradford M, Drazen JM, Hanson B, Howard B, Jamieson KH, Kiermer V, Marcus E, Pope BK, Schekman R, Swaminathan S, Stang PJ, Verma IM. 2018. Transparency in authors' contributions and responsibilities to promote integrity in scientific publication. *Proceedings of the National Academy of Sciences of the United States of America* 201715374. https://doi.org/10.1073/pnas.1715374115. Proposes standards for authorship, describes the responsibilities of corresponding authors and recommends use of ORCID identifiers for authors.

3.10 DECLARE CONFLICTS OF INTEREST

Conflicts of interest include actual, apparent, or potential direct and indirect relationships with individuals and institutions that could influence or bias our work. We must declare any such relationships openly and honestly, for example in peer review and when reporting our work.

3.11 BE KIND

Finally, be kind. Researchers are highly trained critics and are often insecure, so we can be very negative about other people's work. Offer positive and constructive criticism and give praise where it's due. Never criticise a researcher rather than the work. We all benefit from making science inclusive, welcoming, and positive.

3.12 CHAPTER SUMMARY

In this chapter, we've seen that in addition to the ethical requirements we examined in the previous chapter, we must:

- Avoid research misconduct, including fabrication, falsification, and plagiarism.
- Review the literature fairly.
- Use blind protocols where possible.
- Be honest, rigorous, and transparent when conducting science.
- Admit mistakes.
- Use research funds appropriately.
- Respect peer review.
- Assign authorship appropriately.
- Declare conflicts of interest.
- Be kind.

We'll come back to these issues throughout the book. The next chapter looks at inequity, how it damages both people and science and how we can work towards a more inclusive science.

3.13 FURTHER READING

All European Academies. 2017. *The European Code of Conduct for Research Integrity Revised Edition.* Berlin, Germany: ALLEA – All European Academies. www .allea.org/wp-content/uploads/2017/03/ALLEA-European-Code-of-Conduct-for-Research-Integrity-2017-1.pdf [Accessed 3 January 2019]. Defines the principles of research integrity, describes good research practices, and outlines violations of research integrity.

Borries C, Sandel AA, Koenig A, Fernandez-Duque E, Kamilar JM, Amoroso CR, Barton RA, Bray J, Di Fiore A, Gilby IC, Gordon AD, Mundry R, Port M, Powell LE, Pusey AE, Spriggs A, Nunn CL. 2016. Transparency, usability, and reproducibility: Guiding principles for improving comparative databases using primates as examples. *Evolutionary Anthropology* 25: 232–238. Highlights the need to improve the transparency of comparative data and proposes guidelines for doing so.

Center for Open Science: https://cos.io/ [Accessed 3 January 2019]. The Center for Open Science aims to increase openness, integrity, and reproducibility of research. Includes guidelines for transparency and openness in research.

Forstmeier W, Wagenmakers EJ, Parker TH. 2017. Detecting and avoiding likely false-positive findings: a practical guide. *Biological Reviews of the Cambridge Philosophical Society* 92: 1941–1968. https://doi.org/10.1111/brv.12315. Reviews problematic practices in science and highlights strategies to promote better science, including preregistration, blind data collection and analysis, and comprehensive reporting of results.

Fraser H, Parker T, Nakagawa S, Barnett A, Fidler F. Questionable research practices in ecology and evolution. *PLOS ONE* 13: e0200303. https://doi.org/10.1371/journal.pone.0200303. Results of a survey showing that questionable research practices are common in ecology and evolution.

Harris R. 2017. *Rigor Mortis: How Sloppy Science Creates Worthless Cures, Crushes Hope, and Wastes Billions.* New York: Basic Books. Describes the 'reproducibility crisis' in biomedical research; relevant to all science.

Holman L, Head ML, Lanfear R, Jennions MD. 2015. Evidence of experimental bias in the life sciences: Why we need blind data recording. *PLoS Biology* 13: e1002190. https://doi.org/10.1371/journal.pbio.1002190. Reviews how researchers' expectations influence study outcomes and the need for blind protocols.

Ihle M, Winney IS, Krysalli A, Croucher M. 2017. Striving for transparent and credible research: practical guidelines for behavioral ecologists. *Behavioral Ecology* 28: 348–354. https://doi.org/10.1093/beheco/arx003. Outlines the challenges facing science and provides guidelines and tutorials on open practices for behavioural ecologists.

Ioannidis JPA. 2005. Why most published research findings are false. *PLoS Medicine* 2: e124. https://doi.org/10.1371/journal.pmed.0020124. Examines biases that affect research outcomes and how we can address them.

Kerr NL. 1998. HARKing: hypothesizing after the results are known. *Personality and Social Psychology Review* 2: 196–217. Defines HARKing, presents data suggesting that it is common and explains why this is a problem.

Nosek BA, Ebersole CR, DeHaven AC, Mellor DT. 2018. The preregistration revolution. *Proceedings of the National Academy of Sciences of the United States of America* 115: 2600–2606. https://doi.org/10.1073/pnas.1708274114. Distinguishes between postdiction and prediction, and explains why they are often confused. Highlights the benefits of preregistration and addresses some of the practical difficulties involved in implementing preregistration.

Parker TH, Bowman SD, Nakagawa S, Gurevitch J, Mellor DT, Rosenblatt RP, DeHaven AC. 2018. *Tools for Transparency in Ecology and Evolution (TTEE).* http://doi.org/10.17605/OSF.IO/G65CB. A checklist of questions to help authors comply with transparency and openness promotion guidelines and to help reviewers and editors assess that compliance.

Simmons JP, Nelson LD, Simonsohn U. 2011. False-positive psychology: Undisclosed flexibility in data collection and analysis allows presenting anything as significant. *Psychological Science* 22: 1359–1366. https://doi.org/10.1177/0956797611417632. Shows that flexibility in data collection, analysis and reporting dramatically increases false-positive rates and proposes requirements for authors and guidelines for reviewers to resolve this problem.

World Conferences on Research Integrity. *Singapore Statement on Research Integrity.* http://wcrif.org/statement [Accessed 3 January 2019]. An international effort to create a global guide to responsible research conduct.

Smaldino PE, McElreath R. 2016. The natural selection of bad science. *Royal Society Open Science* 3: 160384. https://doi.org/10.1098/rsos.160384. Reviews the incentives that favour poor scientific practice and shows that selection for high output leads to poor science.

Universities UK. *Concordat to Support Research Integrity.* www.universitiesuk.ac.uk/policy-and-analysis/reports/Documents/2012/the-concordat-to-support-research-integrity.pdf [Accessed 3 January 2019]. The UK national framework for good research conduct and its governance.

US Office of Research Integrity: https://ori.hhs.gov [Accessed 3 January 2019]. Useful resources on responsible research conduct and research integrity.

Wicherts JM. 2017. The weak spots in contemporary science (and how to fix them). *Animals* 7: 90. https://doi.org/10.3390/ani7120090. A review of problems in science, possible explanations for them, and how to deal with them.

World Economic Forum Young Scientists. *Code of Ethics for Researchers.* Geneva, Switzerland: World Economic Forum. http://wef.ch/coe [Accessed 3 January 2019]. Seven principles of being an ethical scientist, by an international group of scientists.

4

Inclusive Science

Scientific research is subject to serious inequities of opportunity. Economic, political, social, and cultural influences shape the opportunities available to people. Everyday and institutional practices exclude people based on aspects of their identity. These inequities intersect in complicated ways and have negative effects on both individuals and science. Some may go unnoticed, even by those who are negatively affected by them, because they are so deeply entrenched in our cultures.

In this chapter, I briefly explore discrimination in relation to various aspects of identity, and how these intersect. I then describe the effects of discrimination on people and on science, and how we can help to combat inequities.

4.1 GENDER

Discrimination in relation to gender includes pervasive disadvantages to women in hiring, pay, funding, citations, recommendation letters, and invitations to present in symposia, or to publish in or review for high-impact journals. Women are also under-represented in the laboratories of 'elite' male scientists, and thus miss out on the resources and opportunities associated with training in such laboratories.

Gender issues include childcare and care for other dependents, and are often reduced to this in discussion, but much of gender discrimination is unrelated to parenting.

Both men and women can experience sexual harassment or assault, observe such misconduct, or be informed about it by those who are affected. Perpetrators may be fellow researchers or members of the local community. Surveys suggest that female students and postdoctoral scholars are disproportionately more likely to be on the receiving end of misconduct, and that it tends to be perpetrated by more senior male researchers. Sadly, some senior researchers still don't take this seriously.

4.2 LGBTQIA+

Sexual and gender minorities, including lesbian, gay, bisexual, transgender, queer/questioning, intersex, and asexual (LGBTQIA+) people, face violence, homophobia, and everyday heteronormative assumptions. Trans* people also experience misgendering and exclusion due to the routine use of binary gender categories. Discriminatory legislation and cultural attitudes make some countries extremely dangerous for LGBTQIA+ people.

4.3 ETHNICITY

Racial and ethnic discrimination is an everyday part of many societies. Black and ethnic minority researchers are under-represented in academia. Moreover, people from minority and under-represented groups experience heavy, unacknowledged, and unrewarded workloads associated with roles they are expected to perform.

4.4 SOCIAL BACKGROUND

Science is a historically elite occupation. Social background and socioeconomic status influence access to education and other opportunities. Researchers from low-income or working-class backgrounds, and those scholars whose parents did not attend university (first-generation scholars) face social and cultural barriers, stigma, and the psychological costs of accommodation to a middle- or upper-socioeconomic status environment.

4.5 DISABILITY

People with physical disabilities face practical difficulties as well as prejudice. Less visible or invisible conditions may not be immediately apparent, making them difficult for others to recognise or understand. The discrimination experienced by people living with mental health conditions can exacerbate the difficulties they face. The perception of social stigma can mean that people are less likely to disclose a condition, making them vulnerable to judgement by people unaware of their condition. Moreover, people living with mental health conditions may internalise the social stigma they face, or the perception that such stigma exists, with negative consequences for their self-esteem.

4.6 AGE

Assumptions based on age-related stereotypes affect both the young and the old. For example, young people may be dismissed as inexperienced, naïve, or fragile. Older people can be dismissed as less productive, less energetic, less innovative, or out-of-touch. Some systems actively discriminate against older researchers, imposing age limits on opportunities.

4.7 FIRST LANGUAGE

The hegemony of the English language in science leads to serious disadvantages for scientists whose are not fluent in English. Publications in other languages reach a more limited audience than those in English and may be actively dismissed or simply not be detected in literature searches. For a non-native English speaker, speaking, reading, and writing science in a foreign language requires additional time and energy. Manuscripts by authors from countries where English is the first language are more likely to be accepted for publication, and language barriers can exclude people from discussion.

4.8 COUNTRY OF ORIGIN

Researchers from countries that are under-represented in science may be less familiar, or unfamiliar, with the cultural expectations of the North American and Western European communities that dominate science, leading to marginalisation. Scholars from low- and middle-income countries, including almost all primate-range countries, face a lack of access to research funding, which severely limits their opportunities to conduct research, again leading to marginalisation. Differences in access to training and a lack of available positions lead to serious imbalances in representation in primatology and science more generally.

Stereotypes and prejudice related to country of origin lead to erroneous assumptions about the competence of people from different countries. Researchers from countries that dominate global science may be perceived as, and may perceive themselves to be, superior to researchers from other countries. As a result, the contributions and abilities of scientists from under-represented countries may be under-valued or dismissed, and people may be patronised.

In group discussions, cultural differences can mean that people from some groups contribute more than those from other groups, meaning that important issues are missed.

4.9 INTERSECTIONALITY

Lived experiences are complex and groups defined by single aspects of identity are not homogeneous. We each embody multiple intersecting social identities, which intersect with one another. This **intersectionality** means that one individual may be subject to multiple inequalities, which magnify one another. For example, sexism is compounded by racism for black women and health issues are often compounded by sexism. We may also experience **social privilege** (an advantage available only to a particular group) in relation to one aspect of our identity, yet disadvantage with respect to another. Disadvantage (e.g. being from a low-income background) does not cancel privilege (e.g. being white). Privilege doesn't mean you didn't work hard to get where you are, and it doesn't mean you haven't had a difficult time. It does mean that you had an easier time than people who don't share your privilege.

4.10 THE EFFECTS OF DISCRIMINATION

Discrimination has serious negative effects on both people and our field. Aspects of identity influence where people can live and work safely. Prejudice, discrimination, and social stigma damage and alienate those who experience them. Strategies to avoid harassment and discrimination require energy, placing an additional burden on those who experience them. People from under-represented groups may feel pressure to demonstrate their value and experience imposter syndrome (**Box 1.4**).

Discrimination damages our field because our identity shapes the research questions we ask, how we answer them, and how we interpret our findings. Including the ideas of a diverse group of people thus promotes innovation and creativity and improves science. A lack of role models discourages people from under-represented groups from considering a career in research. By failing to attract a diversity of researchers – by marginalising, patronising, and excluding people based on their identity – and by privileging the English language, we reduce the diversity of researchers, and impoverish science.

4.11 WHAT CAN WE DO TO ADDRESS INEQUITIES?

To help combat the serious inequities in science, we must understand the barriers some primatologists face, both in our own countries and cross-nationally. We must educate ourselves about social privilege and how multiple axes of inequality intersect. We must understand that the actions of those who benefit from a system can perpetuate inequity unless we work against it. Acknowledging that discrimination, bias, and inequity are pervasive can be troubling, as we like to see science as fair and equal, and to think that our own positions are based purely on merit.

We must reflect on our own conscious biases and learn about how unconscious processes help us to make decisions quickly but involve quick judgements that are influenced by our background, cultural environment, and personal experiences, leading to **unconscious bias** (**implicit bias**) in our decision-making. We must understand the influence of subtle and even unconscious gestures, facial expressions, word choice, and tone of voice (**micro-behaviours**) on whether people feel patronised, included, or excluded; and the cumulative effect of subtle discrimination (**micro-inequities**).

We must examine our own behaviour and be aware of power asymmetries. We must acknowledge any privilege we have, recognise the benefits it brings, and use it to promote an inclusive environment. We must work in solidarity with marginalised groups. We must listen to and reflect on feedback and change our behaviour if necessary. We must actively include people and acknowledge contributions fairly. We must support affected colleagues and amplify marginalised voices. We should challenge inappropriate words and behaviours if it is safe to do so and make it clear that we don't tolerate misconduct.

In a discussion, we should seek to hear and understand the viewpoints of everyone involved. If we benefit from a privileged position, we should ask ourselves whether we are dominating a discussion and should listen more than we speak. We should use inclusive language and encourage accurate use of pronouns, for example by stating our own chosen pronouns clearly.

We must accept other people's experiences as valid and believe their accounts. We should not underestimate how difficult it can be to talk about harassment and discrimination. We must understand the cumulative effects of what may seem to be small or even unimportant events. If you think things you are told don't happen, or that people are overreacting, this reflects your privilege.

We should ask ourselves what we can do to make organisations we are part of better reflect the broader population. In multicultural work environments, we should discuss and agree on appropriate behaviour and develop codes of conduct that include how to respond to bullying and discrimination.

We must not allow language to influence our assessment of science. If we speak and write English fluently, we must recognise that this is an advantage, and support those who face a language barrier. If we understand a system, we should share that understanding with others. Creating international, multilingual, and multicultural teams helps to avoid a biased perspective on scientific knowledge.

If we hire staff or volunteers, we should actively encourage applications from all potential qualified candidates, including under-represented groups. We should ensure our processes are fair and transparent. We should develop selection criteria appropriate to the post and base our decisions only on candidates' individual merits in relation to those criteria. If we hire a team, we should seek to include diverse perspectives.

Those of us who mentor students should ensure that we model appropriate behaviour and make it clear that anything else is unacceptable. Senior researchers, in particular, have a duty to openly discuss harassment, bullying, and discrimination, and make it clear that such behaviours are unacceptable, creating an environment in which people feel able to report negative experiences. If we don't actively work to break down barriers to opportunity, we become part of the problem. In other words, silence is complicity.

If you study at or work for an institution, find out what their policies and codes of conduct are and know how to report any incidents. If you plan to work elsewhere, discuss potential study locations, laboratories, and field sites with trusted advisors and peers. Request a code of conduct from potential hosts. If you run a field site, lab, or other workplace, have a code of conduct, and implement it.

If you experience discrimination and your institution doesn't provide adequate support, seek advice from the equity, diversity, and inclusion representatives and committees of national or international learned societies. You are not alone.

If you witness harassment, intervene if you can and help the person who is being harassed to get out of the situation. If you cannot intervene, support the harassed person by corroborating their account of what happened.

4.12 CHAPTER SUMMARY

In this chapter, we've seen that:

- Aspects of an individual's identity can lead to serious inequalities of opportunity.
- Groups are not homogeneous and social identities intersect, meaning that one individual may be subject to multiple inequalities, which compound one another.
- Harassment and discrimination have negative effects on the people who experience them, and also damage science by reducing the diversity of perspectives we include.
- We can help to combat discrimination and work towards creating an equal-opportunity global scientific community by understanding social privilege and marginalisation and actively promoting inclusive science.

The next chapter looks at another fundamental topic in primatological research: how we use statistical analysis to test whether our observations support our predictions.

4.13 FURTHER READING

Amano T, González-Varo JP, Sutherland WJ. 2016. Languages are still a major barrier to global science. *PLoS Biology* 14: e2000933. https://doi.org/10.1371/journal.pbio.2000933. Reviews the potential consequences of language barriers for science and proposes practical solutions.

Anon. 2016. Is science only for the rich? *Nature* 537: 466–470. https://doi.org/10.1038/537466a. Reports on how poverty and social background remain huge barriers in scientific careers, with examples from eight countries around the world.

Bicca-Marques JC. 2016. Development of primatology in habitat countries: A view from Brazil. *American Anthropologist* 118: 140–141. https://doi.org/10.1111/aman.12503. Reviews key events in the establishment and consolidation of Brazilian primatology and strategies to empower national scholars and advance the field of primatology in habitat countries.

Clancy KBH, Nelson RG, Rutherford JN, Hinde K. 2014. Survey of academic field experiences (SAFE): Trainees report harassment and assault. *PLOS ONE* 9: e102172. https://doi.org/10.1371/journal.pone.0102172. Reports the results of a survey of field scientists' experiences of sexual harassment and sexual assault, showing that trainee women are particularly affected.

Fan P-F, Ma C. 2018. Extant primates and development of primatology in China: Publications, student training, and funding. *Zoological Research* 39: 249–254. https://doi.org/10.24272/j.issn.2095-8137.2018.033. Reviews the development of primatology in China, and strategies to promote further development.

Freeman J. 2018. LGBTQ scientists are still left out. *Nature* 559: 27–28. https://doi .org/10.1038/d41586–018-05587-y. On heteronormative assumptions and bias that disadvantage scientists from sexual and gender minorities.

Gutiérrez y Muhs G, Niemann YF, González CG, Harris AP. 2012. *Presumed Incompetent: The Intersections of Race and Class for Women in Academia*. Logan: Utah State University Press. Essays on the challenges faced by women of colour in US academia.

Hoàng TM. 2016. Development of primatology and primate conservation in Vietnam: Challenges and prospects. *American Anthropologist* 118: 130–137. https://doi.org/10.1111/aman.12515. An evaluation of the development and future of primatology in Vietnam.

Iyer N, with contributions from Lutz M, McInturf A, Lau A. 2018. Beyond the scientific bubble: The inequity dilemma in field research. *The Ethogram*. https://theethogram.com/2018/02/15/beyond-the-scientific-bubble-the-inequity-dilemma-in-field-research/ [Accessed 3 January 2019]. An excellent blog post on inequities in field research.

Marín-Spiotta E 2018. Harassment should count as scientific misconduct. *Nature* 557: 141. https://doi.org/10.1038/d41586–018-05076-2. Argues that scientific integrity should include how we treat people, as well as how we handle data.

Mehta D. 2018. Lab heads should learn to talk about racism. *Nature* 559: 153. https://doi.org/10.1038/d41586–018-05646-4. Calls on senior academics to lead in discussions of intolerance.

Nelson RG, Rutherford JN, Hinde K, Clancy KBH. 2017. Signaling safety: Characterizing fieldwork experiences and their implications for career trajectories. *American Anthropologist* 119: 710–722. https://doi.org/doi:10.1111/aman .12929. Assesses the effects of experiencing gender-based discrimination, harassment, and assault on field researchers and ways to improve field experiences and achieve equality of opportunity.

Savonick D, Davidson CN. Gender bias in academe: An annotated bibliography of recent studies of academic gender bias and gender discrimination. The Impact Blog. London School of Economics and Political Science. http:// blogs.lse.ac.uk/impactofsocialsciences/2016/03/08/gender-bias-in-academe-an-annotated-bibliography/ [Accessed 3 January 2019]. A very useful collection of studies of gender bias in the academy, with talking points and summaries.

Setchell JM, Gordon A. 2018. Editorial practice at the International Journal of Primatology: The roles of gender and country of affiliation in participation in scientific publication. *International Journal of Primatology* 39: 969–986. https:// doi.org/10.1007/s10764–018-0067-1. Investigation of editorial practices at the *International Journal of Primatology* with respect to gender and country of affiliation.

Timesup: www.timesupnow.com [Accessed 3 January 2019]. A US-focussed website with resources about sexual assault, harassment, and inequality in the workplace.

5

Understanding Statistical Evidence

Statistical evidence is fundamental to science. Understanding statistics helps us to understand the literature and assess it critically, refine our research questions into testable hypotheses and predictions, design studies that are appropriate to test these predictions, evaluate whether our findings support our predictions, and derive appropriate conclusions. The dominant paradigm in primatology and allied disciplines is to test whether patterns we perceive in our observations are due to more than random variation in our data. However, the statistical analyses we use to do this are very often misinterpreted.

In this chapter, I introduce key concepts in statistics. I begin by explaining how we use statistical analysis to infer something about a theoretical population based on a sample. I distinguish between different kinds of variables, then introduce relationships between them I introduce null hypothesis significance testing, the most widely used approach to statistical testing in primatology, and explain common misunderstandings of this approach. I review the two types of error that arise in null hypothesis statistical testing and the concept of statistical power. I explain the need to assess and report effect sizes and confidence intervals as well as statistical significance, then briefly introduce alternatives to null hypothesis statistical testing. I end with how to interpret statistical results appropriately, in the context of other evidence.

5.1 INFERRING FROM A SAMPLE TO A POPULATION

As scientists, we seek to answer general questions about a set of cases, for example a set of individual animals, groups, or species. We term this set a **population**. Statistical populations are different from the everyday understanding of population and can be quite abstract. For example, a population might be all the individuals of a species that we could theoretically measure, including those that have lived, are alive, and will live in future.

The properties of populations are called **parameters**, and they don't change because they are calculated for the entire population. Parameters include proportions, means, and standard deviation (Box 5.1). For example, the proportion of females in a population might be 50%.

If we could measure all the cases in a population, we would be able to measure parameters perfectly. However, that's not usually possible, so instead we measure the properties of part of the population (a **sample**) and use these properties to draw conclusions about the population. We term this **inference**, because there is obviously some uncertainty associated with drawing conclusions about a population based on a sample.

The properties of a sample are termed **statistics**. Unlike parameters, statistics vary, because they are calculated for only some of the population. If we measure a different sample, we will get a different value. In our example, we might find a lower or higher proportion of females in our sample than in the overall population, depending on the sample.

The size of our sample matters in terms of how accurate our estimates of population parameters are. If we sample just a few individual cases from a population, our estimates of population parameters are less likely to be accurate than they are if we sample a large number of cases. For example, the mean body mass of an adult female of a primate species might be 12.5 kg in one sample, 10.8 kg in another, and 12.3 kg in a third, even if the samples are all drawn from the same population. We call this random variability in a statistic across samples **sampling error**.[1]

Sampling error means that samples don't represent a population perfectly, so we use **inferential statistics** to assess the statistics for a sample and draw general conclusions (inferences) about the corresponding population parameters.

Sample size affects the reliability of a study. Published studies based on small sample sizes often have larger or more variable effects than those based on larger samples, due to random noise in the data and publication bias.

If we wish to generalise the statistics we derive from our sample to the population, the sample must be **representative** of the population.

---

[1] Error in this context doesn't mean that we made a mistake.

Box 5.1: Descriptive statistics

Unless we have very few observations, we can't present raw data to an audience. Instead, we summarise a set of observations using **descriptive statistics**. These include measures of **central tendency** and the dispersion or **spread** of the data around that central value.

Central Tendency

The central tendency of a set of observations is a single value that represents the centre point or typical value in that dataset (the **average**).

The arithmetic **mean** (often just termed the mean) of a set of observations is the sum of all the values, divided by the number of observations (**sample size, N**). The mean is the most common measure of central tendency. It is influenced by **outliers** (values that are much greater or smaller than most of the values). In skewed distributions, the mean is not the same as the middle value (**median**), or the most common value (**mode**).

Always specify the type of central tendency you describe. In other words, use *mean*, *median*, or *mode*, not *average*.

Spread

Always report a measure of the spread of data around the central tendency. All the following measures have the same units as the data.

For small datasets, we simply report the **range** – the smallest and the largest values.

We can also report the **inter-quartile range**, or middle 50%, by dividing the data into four equal parts (quartiles). The values that separate the first from the second and the third from the fourth quartiles are the inter-quartile range and are often plotted in **box-plots** with the median (which is the second quartile).

For large datasets, we have three common choices, which we report as +/− or ± in text and as error bars in plots.

The **standard deviation (SD)** measures the spread of the individual data points in a set of observations. Approximately 68.27% of observed values lie within one SD of the mean, and approximately 95.45% of observed values lie within two SDs of the mean. We use the SD to describe variability in a sample.

The **standard error** of the mean (**SEM**) describes the accuracy with which the sample mean represents the population mean. It

---

Box 5.1: (cont.)

measures how much you would expect the mean to vary if you repeat your data collection many times using the same sample size. In precise terms, it is the expected SD of the distribution of the means. The SEM is the SD divided by the square root of the sample size (SEM $= $ SD$/\sqrt{N}$) and decreases with the sample size.

Confidence intervals (CIs) describe the level of uncertainty in your estimate of a parameter (i.e. how precise it is). CIs define the boundaries within which most estimates of a parameter will fall if you repeat data collection many times.

Measures of spread are commonly misused. Always check which measure of spread around the central tendency a report uses, and when writing always state the measure you use. Authors may choose to report the SEM because it is smaller than the SD and makes small differences between groups appear larger to the uncritical eye. Conversely, they may use SD in an attempt to distract the reader's attention from differences between groups. Don't be tempted to do this.

When reporting your own data, use the SD to describe variability in the data, and appropriate CIs (often 95%, but decide this for yourself) to indicate the degree of uncertainty in an estimate.

Finally, the **coefficient of variation (CV)** is the ratio of the SD to the mean, expressed as a percentage. It has no units. It is useful to describe the relative variation in a sample or population. It is only useful for data with positive values and is not useful where the mean value is zero or very close to zero.

In other words, all the cases in a population must be equally likely to be selected for the sample. If some cases in the population are more likely to be included than others, this results in a **biased** sample. The extent to which we can justify extending our conclusions from the results of a study based on a sample to a broader population (i.e. how much we can generalise from our findings) depends on how representative that sample is of the population.

5.2 VARIABLES AND DISTRIBUTIONS

A **variable** is a quality or quantity that varies. There are different ways to classify variables. In this book, I distinguish between categorical, ordinal, and continuous variables because the differences between these types of data have important implications for statistical analysis.

Categorical variables (or **nominal variables**) are qualitative labels with no numerical significance. Examples are behavioural categories (e.g. feeding, resting, travelling), blood groups, and types of bone. Categories are mutually exclusive; an individual case can only fall into one category. We can record counts and determine the mode (Box 5.1), but categorical variables have no order, so doing any sort of calculation with the values would be meaningless. A **binary variable** is a special case of a categorical variable with only two possible values (e.g. present vs. absent).

Ordinal variables are quantitative. They have a defined order but no information about the differences between values. An example is traditional dominance ranks, which sort animals into an order, but are mute on the relative difference between animals 1 and 2 and animals 2 and 3. We can calculate the mode and the median (Box 5.1).

Categorical and ordinal variables can take only particular values; in other words, they are **discrete**. In contrast, **continuous variables** can take any value in a range. They have an order and consistent differences between values. Examples are age, body mass, time, and temperature. We can calculate the mean, median, and mode (Box 5.1). Continuous variables are divided into interval and ratio measures, depending on whether zero has a clear definition, but most of the time we treat them in the same way.

The frequency **distribution** of a dataset is all possible values of the data and how often they occur. We plot data as a **histogram** with the scores on the x-axis and the frequency of each score on the y-axis. Distributions take many forms and underlie our choice of statistical analyses. Many datasets approximate a symmetrical bell-shaped curve clustered around a central value. This **normal distribution** (or **Gaussian distribution**) is widely used in statistics.

5.3 STATISTICAL RELATIONSHIPS BETWEEN VARIABLES

Most of our predictions in primatology concern statistical relationships between variables. These relationships take two basic forms: differences between groups and associations between quantitative (ordinal and continuous) variables, where change in one variable is linked to change in the other.

We are often interested in an association between two variables because we hypothesise that change in one of the variables *causes* change in the other variable. In such situations, the variable that we think is the cause is the **predictor variable** (**independent variable** or **explanatory variable**), and the variable that we think is the effect is the **outcome variable** (**dependent variable** or **response variable**).[2] By convention, we plot the predictor variable on the x-axis and the outcome variable on the y-axis.

If we find a relationship between two variables (a **correlation**), we cannot necessarily conclude that they are meaningfully related to one another, or that change in one variable causes change in the other (**causation**) for three reasons. First, a relationship may be due to chance. Second, causality may be the reverse of what we assume. In other words, B causes A rather than A causing B (the **directionality problem**). Third, the two variables may be unrelated, but change in a third variable (a **confound** or **confounding variable**) may cause change in both variables, causing a spurious relationship. For example, the number of times we observe a particular behaviour will correlate with the number of times we observe another behaviour, if we don't account for the time we observe each individual, simply because we see more of all types of behaviour if we observe an individual for longer (i.e. time observed is a confound). Similarly, the number of offspring a female has is related to tooth wear, but only because both are related to age (i.e. age is a confound).

5.4 NULL HYPOTHESIS STATISTICAL TESTING

The conventional (or **frequentist**) approach to statistical testing is to test the **null hypothesis** that there is no difference between groups or no relationship between variables. We do this by calculating a numerical summary of our sample data (a **test statistic**) and comparing it with a predetermined threshold value (a **critical value**) to draw conclusions about whether we can reject the null hypothesis.

If our statistical analyses reject the null hypothesis, we interpret this as evidence in favour of an **alternative hypothesis**, that there is a difference between groups or a relationship between variables.

Although null hypothesis statistical testing pervades research in primatology and other disciplines, it is severely criticised and very often

---

[2] I use predictor and outcome variable throughout this book, for consistency.

misunderstood. One major criticism is that null hypothesis statistical testing tells us how likely our observed result is, given that the null hypothesis is true. However, we usually want to know the inverse of this – the probability that the hypothesis is true, given the data we have. In other words, null hypothesis statistical tests don't tell us what we want to know.

A second criticism is that a rejected null hypothesis doesn't really say anything about the alternative hypothesis (our prediction). It only means that the data are unlikely to have emerged from the process assumed under the null hypothesis. If those data are limited or unclear, the conclusions will also be ambiguous. In other words, null hypothesis testing doesn't allow us to falsify our research hypothesis.

A third criticism is that null hypothesis statistical testing encourages dichotomous thinking (yes/no answers), such as whether there is a difference between groups, or a relationship between variables. However, the more interesting question is often how large a difference there is between groups, or how strong a relationship is.

A fourth criticism is that a focus on null hypothesis statistical testing distracts attention from the aim of the research. Valuable descriptive data are often drowned by unnecessary null hypothesis testing, often without good justification. In some cases, authors don't even present the actual data.

Finally, a focus on statistical significance, rather than effect size, underlies many of the questionable research practices we reviewed in Section 3.5, including *p*-hacking and HARKing.

5.5 WHAT THE *p* VALUE MEANS AND DOESN'T MEAN

In null hypothesis statistical testing, we calculate a test statistic and compare this with the expected distribution of the test statistic under the null hypothesis (a **statistical model**). All such tests are based on **assumptions** about the population from which we obtained our sample. The probability value (**p value**) is the probability that the test statistic would have been as large, or larger, than the observed value if the null hypothesis is true and the assumptions underlying our test hold.

We can understand the *p* value as a measure of the fit of the statistical model to the data. The smaller the *p* value, the stronger the evidence against the model (including the null hypothesis). The

p value ranges from 0, where the model is completely incompatible with the data, to 1, where the model is perfectly compatible with the data. If the p value is less than a threshold value, we reject the null hypothesis and declare the result statistically significant. This threshold is often, but arbitrarily, set at 0.05 (5%).

Statistical packages calculate test statistics and p values for us, but they don't do the thinking for us, so we must understand how the test works to avoid making mistakes.

The p value is very often misunderstood and misused.

The p value DOES NOT measure the probability that the data were produced by chance. In other words, the p value does not measure the probability that the null hypothesis is true. Instead, the p value tells us the probability of obtaining the data we collected, assuming that the null hypothesis is true. This is unlikely to be what we want to know.

Similarly, **$1 - p$ is NOT the probability that the alternative hypothesis is true.** It is tempting to think that if $p = 0.02$, then the probability that the alternative hypothesis is true is 0.98, or 98%. This is not true. The calculation of the p value is based on the assumption that the null hypothesis is true. The p value can't tell us whether this assumption is correct.

The p value DOES NOT measure the size or importance of the effect we observe. This is a common misconception, possibly because *significant* has a general meaning of *important*.

We can obtain a very small p value either by measuring a very large effect or by measuring a small effect with great certainty. With a sufficient sample size, you can detect a significant but very tiny effect and reject the null hypothesis. Conversely, with a small sample size, a large and important difference may be statistically non-significant. In other words, the same **effect size** is associated with a different p value depending on the sample size, and the same p value can be associated with different effect sizes, depending on the sample size.

You may read studies that claim that increasing the sample size would lead to a significant result. However, this is circular reasoning. Increasing the sample size will only lead to a significant result *if* the null hypothesis is false, which is what we wish to test.

A non-significant p value DOES NOT mean there is no difference between groups or no relationship between two variables. It is simply inconclusive. Researchers often interpret a non-significant result as a lack of effect (i.e. that the effect is 0). However, we never accept the null hypothesis. In other words, absence of evidence is not evidence of absence. A non-significant test tells us only that the difference we observe may be due to random variation in the dataset, or may be a real effect, but that the sample is too small to detect it.

A *p* value is a poor predictor of future *p* values. Replications of the same experiment can result in quite different *p* values.

5.6 FALSE POSITIVES, FALSE NEGATIVES, AND STATISTICAL POWER

Null hypothesis significance testing can lead to two types of error.

False positives (Type I errors) occur when you reject the null hypothesis, but it is actually true. The risk of committing this error, α, is very often set to 0.05. This is the level of significance for your test. Using a lower α decreases the risk of a false positive but increases the chance of failing to detect a true difference.

False negatives (Type II errors) occur when the null hypothesis is false, but you fail to reject it. The probability of a false negative, β, depends on the probability that a test will reject the null hypothesis when the alternative hypothesis is true (**statistical power**, $1 - \beta$). In other words, statistical power is the probability that a study will detect an effect when an effect exists. You can decrease the risk of a false negative by increasing the statistical power.

If we run multiple null hypothesis statistical tests, the probability of a false positive increases. For example, if we run 20 tests, with a significance threshold of 0.05, we have a 64% chance of observing a significant result by chance, in the absence of a real effect.[3] This problem of **multiple testing**, or **multiple comparisons**, underlies many of the problems we reviewed in Section 3.5, if we fail to report all the tests we conduct. We can avoid this by running a small number of planned comparisons, correcting for multiple testing by using a lower value of α, and reporting our analyses honestly.

Traditional null hypothesis significance testing focusses heavily on avoiding false positives, and almost ignores the risk of false negatives. However, the two errors trade off- as α increases, β decreases- so we need to take both into account when designing studies and analysing data. We'll go into more detail on this in Chapter 16. For now, bear in mind that many studies in primatology are under-powered, leading to an increased risk of both false negatives and false positives.

---

[3] The probability of at least one significant result is calculated as $1 - p$ (no significant results) $= 1 - (1 - 0.05)^{20} = 0.64$.

5.7 EFFECT SIZES

As we have seen, testing for significance does not tell us anything about the magnitude of an observed effect.[4] The effect size is either the size of the difference between two groups (e.g. males and females) or a measure of the degree of association between two variables (e.g. between brain size and body size).

There are statistical conventions for standardising effect sizes, including *r*, *Cohen's D*, and *odds ratios*, but it is more useful to think of them in terms of what we can measure in the real world. In other words, what does the result mean? For example, are males twice the size of females, or 2% larger than females? Does body size explain all variation in brain size, or a very small percentage of it?

Reporting effect sizes separates the strength of the real-world effect from whether the findings are likely to be due to chance.

We calculate an effect size for a sample, to represent that of the population. An effect size based on a large sample is likely to be more accurate than one calculated from a small sample.

5.8 CONFIDENCE INTERVALS

We should also present the degree of confidence we have in our estimated effect size (i.e. how precise it is). We can use confidence intervals to do this (Box 5.1). A narrow confidence interval tells us that we can be confident in our estimate. A wide interval tells us that the measurement is imprecise.

We can state CIs at any level, but 95% is common. A 95% confidence interval means that if we sampled the same population repeatedly and estimated the confidence interval each time, 95% of those intervals would contain the true population parameter. It does not mean that there is a 95% probability that the interval contains the true population parameter. This is a common misinterpretation.

If a confidence interval does not cover a hypothesised parameter value, then our data do not support that value. If the confidence interval includes a value, we cannot reject that value. If a 95% confidence interval describing an effect includes zero, the effect is not significantly different from zero (at $\alpha = 0.05$). In other words, CIs can answer the same questions as *p* values. However, they provide more information,

---

[4] Beware: 'effect' in this context does not necessarily mean causality.

because they include information about the precision of the estimated parameter. Large samples increase precision and decrease the size of the confidence interval.

You may also hear that if the 95% CIs of two means overlap, then the means are not significantly different (at $\alpha = 0.05$), or that if the mean of one group is outside the 95% confidence interval of another group, the means are significantly different. Neither of these statements is true. If we want to know whether two means are significantly different, we should use an appropriate statistical test (more on those in Chapter 15).

Examining the effect sizes reported in the literature is much more informative than focussing on the results of null hypothesis testing. Rather than contrasting studies that did and didn't find a statistically significant effect, based on an arbitrary threshold, we should focus on the size and precision of reported effects, and how much they overlap.

5.9 ALTERNATIVES TO NULL HYPOTHESIS STATISTICAL TESTING

Alternatives to null hypothesis statistical testing weigh the support for different scenarios, rather than evaluating how likely an observed result is if we assume that the null hypothesis is true. This means that they are often more closely aligned to the questions we really want to know the answer to than null hypothesis statistical testing is.

Model selection involves comparing the support for several hypotheses (models) and identifying one or more models that are most appropriate for the data. Various criteria exist to select models that combine the best fit to the observed data with minimal complexity. Without careful formulation of hypotheses, model selection is exploratory, not confirmatory (Section 3.5).

Bayesian inference allows us to incorporate prior information into our analysis and estimate the probability that a hypothesis is true. In other words, we learn from data rather than repeatedly testing the null hypothesis. This is more intuitive than the assumptions underlying null hypothesis statistical testing. Bayesian inference is relatively uncommon in primatology but has great potential in our field, allowing us to concentrate on quantifying the most probable effect sizes, in the light of theory, previous studies, and the quality and limitations of a study.

5.10 INTERPRETING STATISTICAL EVIDENCE APPROPRIATELY

Now that we know what the p value means and what it doesn't mean, and understand types of error, effect sizes, and CIs, we can interpret p values appropriately. To do so, we should use the p value as one piece of a puzzle, along with existing scientific evidence, the plausibility of the proposed explanation, and the quality of the study design. We should examine the effect size and CIs and the exact p value (not just whether it is above or below a threshold). We should consider the assumptions underlying a test carefully, and the power of the test. We should remember that a statistical hypothesis describes patterns in data, there are many reasons why a relationship might occur and the same prediction can support multiple hypotheses. In other words, we can't draw conclusions about a scientific hypothesis from a statistical hypothesis test alone. Finally, as we saw in Section 3.5, an inappropriate focus on statistical significance and p values leads to the questionable research practices such as p-hacking. To avoid these, we should report studies honestly, including all the statistical tests we conduct, and distinguish clearly between exploratory and confirmatory research (Section 3.5).

5.11 CHAPTER SUMMARY

In this chapter, we've seen that:

- Understanding statistics helps us to understand and assess the literature, and design and conduct our own study.
- We use inferential statistics to infer something about a statistical population based on a representative sample drawn from that population.
- We use statistical analysis to test whether our observations support our predictions.
- Predictions take two basic forms: differences between categories and associations between quantitative variables.
- Correlation does not mean causation.
- Null hypothesis statistical testing is commonly used and commonly misunderstood.
- p values are also commonly misunderstood and do not indicate the magnitude of an effect.

- We should examine real-world effect sizes and how confident we are in them as well as the results of null hypothesis significance tests and p values.
- Alternatives to null hypothesis testing include model selection and Bayesian inference.
- We must interpret statistical evidence in the light of existing knowledge, study design, and plausibility.

We'll come back to statistical analysis in more detail later on in Chapter 15. Meanwhile, the next chapter looks at the essential skill of writing.

5.12 FURTHER READING

Cumming G, Calin-Jageman R. 2012. *Understanding the New Statistics: Estimation, Open Science, and Beyond.* New York: Routledge. Covers how null hypothesis statistical testing leads to dichotomous thinking and the advantages of estimation based on effect sizes, CIs, and meta-analysis. 'New' in the title means 'new to most researchers' rather than that the methods are new.

Ellis PD. 2010. *The Essential Guide to Effect Sizes: Statistical Power, Meta-Analysis, and the Interpretation of Research Results.* Cambridge: Cambridge University Press. A simple guide to reporting and interpreting effect sizes. Written for the social sciences, but just as useful for primatology.

Greenland S, Senn SJ, Rothman KJ, Carlin JB, Poole C, Goodman SN, Altman DG. 2016. Statistical tests, P values, confidence intervals, and power: A guide to misinterpretations. *European Journal of Epidemiology* 31: 337–350. https://doi.org/10.1007/s10654–016-0149-3. A very useful explanation of the misinterpretation of statistical tests, p values, and confidence intervals.

Loftus GR, Masson MEJ. 1994. Using confidence intervals within subject designs. *Psychonomic Bulletin & Review* 1: 476–490. https://doi.org/10.3758/bf03210951. Begins with the history of hypothesis-testing and goes on to review the use of CIs in figures to help the reader assess patterns in the data.

Reinhart A. 2015. *Statistics Done Wrong: The Woefully Complete Guide.* San Francisco, CA: No Starch Press. A highly readable guide to practicing statistics responsibly. Clear and enjoyable.

Smith RJ. 2018. The continuing misuse of null hypothesis statistical testing in biological anthropology. *American Journal of Physical Anthropology* 166: 236–245. https://doi.org/10.1002/ajpa.23399. Explains how null hypothesis statistical tests are misinterpreted in biological anthropology and why we should use effect sizes and CIs.

Vigen T. 2015. *Spurious Correlations.* New York: Hachette Books. Humorous examples of spurious correlations in public datasets. Also see www.tylervigen.com/spurious-correlations.

Wasserstein RL, Lazar NA. 2016. The ASA's statement on p-values: context, process, and purpose. *The American Statistician* 70: 129–133. https://doi.org/10.1080/00031305.2016.1154108.

6

Communicating Ideas in Writing

The ability to write is an essential component of research. We write to communicate with readers. Our readers include funding bodies, thesis examiners, manuscript editors, reviewers, and readers of a journal. In each case, we write to convince a reader of our argument. In reports, we also write to allow a reader to check and interpret our findings for themselves.

Good writing conveys information to readers as clearly and simply as possible. Poor writing obscures meaning, frustrates the audience, and puts them off reading our work. Poorly crafted writing can make the reader suspect that our science may also be confused.

In this chapter I cover general points, that apply to all scientific writing. I begin with advice on drafting, and the need to revise, obtain feedback, and revise your draft again. This iterative process can come as a surprise to students accustomed to submitting work for a deadline, then forgetting about it. I then cover general style, followed by specific topics. I concentrate on writing for a scientific audience but cover writing for a general readership briefly in Box 6.1.

6.1 START BY WRITING

Most, if not all, good writing starts out as bad writing. The difference is editing. To write well, begin by writing; then edit what you've written. Your first draft is never your final draft so don't worry about good writing at this stage.

Start writing early; don't procrastinate. You need time to write, edit your own work, seek feedback, and act on that feedback. Writing clarifies your thinking and helps resolve intellectual confusion. Concept maps, arranging sticky notes, and similar methods help to organise your thoughts, as does explaining your work to someone else. Word

Box 6.1: Writing for a non-scientific audience

There are many reasons to write for a non-scientific audience. Some funding organisations and scientific journals require lay (non-technical) summaries of a project. We report to stakeholders, decision-makers, funding agencies, and host institutions. We also have a responsibility to communicate what we discover to the public.

Science communication is a skill. It's not easy and may not come naturally. As with all writing, focus on communicating with your audience. Practice and enthusiasm help, and feedback from non-expert friends and colleagues will reveal what works and what doesn't work. Remind yourself of common misunderstandings about science (Box 1.1).

Decide on the point you wish to make and communicate this message concisely and clearly. Get your point across quickly, then give the broader context of what we know about a topic (the opposite of a scientific report). Remember that what is familiar to you may not be familiar to your reader. Don't overdo the detail. Be accurate. Don't exaggerate. Interpret effect sizes for the reader in real-world terms. Use examples, analogies, and metaphors to help the reader relate to concepts. If you're reporting to stakeholders, make pragmatic and clear recommendations, based on a summary of the evidence.

Use accessible language and translate technical terms into plain language. Jargon and unnecessarily complicated words can confirm people's stereotypes about scientists and reinforce barriers between the general public and researchers. It's easy to forget that much of what researchers consider to be everyday vocabulary is unfamiliar to most people. Moreover, words we commonly use in science have different meanings in everyday language and can cause confusion. For example, we use *significant* to mean statistical significance, but it means *important* to a general audience, so write about whether patterns might be due to *chance* or how *confident* we are in our results. To a scientist, a *positive* trend is a positive relationship between variables, but to a general audience it's a *good* trend, so use *upward* to avoid confusion, or explain that x increases with y. To a scientist, *error* is the difference between a measure and the real value; to a general audience an *error* is a *mistake,* so rephrase.

Box 6.1: (cont.)

Never publish photographs of you with your study species on social media, on your website, in reports, or elsewhere. Doing so encourages viewers to think that primates are suitable as pets. They are not (Section 2.11).

FURTHER READING

Barron AD, Brown MJ. 2012. Science journalism: Let's talk about sex. *Nature* 488: 151–152. https://doi.org/10.1038/488151a. Explains that maintaining a consistent and objective message when communicating with journalists can improve the way in which our results are covered.

Salita JT. 2015. Writing for lay audiences: a challenge for scientists. *Medical Writing* 24: 183–188. https://doi.org/10.1179/2047480615Z.000000000320. Includes a useful table of terms that have different meanings for lay audiences.

Sumner P, Vivian-Griffiths S, Boivin J, Williams A, Venetis CA, Davies A, Ogden J, Whelan L, Hughes B, Dalton B, Boy F, Chambers CD. 2014. The association between exaggeration in health related science news and academic press releases: Retrospective observational study. *British Medical Journal* 349: g7015. https://doi.org/10.1136/bmj.g7015. Shows that exaggeration in news coverage is linked to exaggeration in press releases and that we have the opportunity to improve reporting by improving press releases.

There's lots of advice on the Internet, including a de-jargoniser: http://scienceandpublic.com. The first paragraph of this box is 95% suitable for a general audience.

The Up-Goer five challenge uses Theo Sanderson's text editor (http://splasho.com/upgoer5/) and is based on webcomic *xkcd*'s description of a space rocket using only the 1000 most used words in English (https://xkcd.com/1133/).

processors have a useful outline view which allows you to see the structure of your text and reorder sections easily.

Minimise distractions and identify your natural habits when writing. For example, I write well in the morning, not in the afternoon, and well again in the evening. Some people work better in the library or in cafes, where they are removed from easy distractions like household chores and online entertainment. Work with the patterns you identify. Take advice from more experienced colleagues about what works for them and try it for yourself.

Experiment with writing strategies. For example, you might find it works well to write notes, order these into a draft, then go back and fill in the detail. A related possibility is to start with headings and bullet points for the main ideas you wish to communicate.

Start writing and keep going. Leave space and move on rather than stopping to look up a citation or finish a sentence or paragraph that you're stuck on. Use colour, highlights, or insert '????' so you can find the place again later. Make notes for yourself in a different font or colour. Use informal colloquial language in your first draft, then redraft in formal language later, if that helps you to write.

Don't stop at the end of a section when writing. Instead, begin the next one before stopping. This makes it easier to start your next session. Another way to ease yourself back into writing is to start filling in the blanks you left earlier.

6.2 REFINE YOUR DRAFT

Your writing won't be perfect straight away. Once you have a draft, put it aside for at least two days. Plan for this in your timeline.

When you come back to your draft you will have fresh eyes and be ready to refine it. Save the original file and make a copy to work on. Are the order and flow of your ideas logical and easy to follow? Check your reasoning and look for gaps. You are familiar with the topic, but your reader may be less so. Simplify and clarify the text. Read the text aloud and check for phrases that trip you up. Is there a better way to present your ideas? Check for ambiguity Might your reader misunderstand you? Cut material you don't need. This can be hard, so paste it into a different document if you can't bring yourself to delete it.

Check any formatting instructions you have and follow them. For example, scientific journals have specific formatting styles.

6.3 GET FEEDBACK AND REVISE YOUR DRAFT

Once you have a good draft, and have refined it yourself at least once, ask someone else to read it. This is the first test of whether you have communicated your ideas clearly and can also generate new ideas. Provide your reader with specific questions that you would like feedback on.

Ask experts in your field to check the details of your draft and people from other fields to check that your arguments are clear. If you are a non-native English speaker, get someone to check your English (Box 6.2). Ideally, ask a native English-speaking colleague in the same

Box 6.2: What if English is not my first language

There are three main options for writing in English if it is not your first language:

1. Write in English from the first draft. This is by far the best way. It takes effort and practice and, yes, it's difficult, but ultimately you will read and write the language of science fluently.
2. Write in your own language, and then translate it. This takes longer and leads to awkward phrasing and style. (Online translation tools are not useful for scientific English and can produce gibberish.)
3. Write in your own language and use a professional translator. This can be expensive, often results in errors (e.g. use of *children* for *juvenile primates*), and rarely results in well-written text.

Reading helps you to identify and absorb good practice. Seek out well-written articles and learn from them. Seek advice and feedback and double-check translations with other people. Learn from comments and generalise from them. Make a list of errors you often make and check for them in your own drafts. Use grammars and style guides, and turn on the spelling and grammar checkers in your word processing software. Apostrophes and *the* cause particular problems.

Always use your own words. Don't be tempted to copy text by another author, even into your notes, because you may forget that you didn't write it, and use it later as though you wrote it yourself. This is plagiarism and will get you into serious trouble (Section 3.1). Quoting text without attribution may be culturally acceptable, or even encouraged, in some countries, but it is unacceptable in scientific reports.

Look out for **false friends** (deceptive cognates). These are words that look or sound similar but differ subtly, or even radically, in meaning between languages. For example:

1. *actually* means *in fact* in English but similar words in other languages mean *currently*
2. *eventually* means *in the end* in English but similar words in other languages mean *possibly*
3. *important* in French does not always translate to *important* in English

Box 6.2: (cont.)

4.   *realise* usually means *become aware* in English, but *realiser* means *to carry out* in French and *realizar* means *to make* in Spanish
5.   *proper* means *seemly* in English as well as *correct*, so use *appropriate*
6.   *so-called* implies doubt in English
7.   we *review* other people's work, they *revise* it
8.   we *record* behaviour, we don't *register* it

There are many other examples, so make a list and learn them.

FURTHER READING

Glasman-Deal H. 2009. *Science Research Writing for Non-Native Speakers of English*. London: Imperial College Press. Dissects each section of a standard report, with grammar and vocabulary. Aimed at authors with English language ability of intermediate level or above and also useful for native speakers who are not confident in their writing. Recommends some things I don't agree with, like use of the passive voice, and including methods and discussion in the results.

field to read your work. They will be able to understand what you were trying to say and to suggest ways to convey what you meant.

Seeking feedback makes version control very important. Be systematic and organised. Remember that your readers and co-authors will need time to read and think about your writing. Agree on a reasonable timeline with them and prompt them if they do not provide feedback when agreed.

Listen to, or read, feedback carefully. Don't simply rebut the points your reader makes. If one reader doesn't understand your text, others won't either, so revise it. Generalise from the advice you receive and learn from it.

Save drafts of your work and keep earlier versions in case a file is corrupted. This is particularly likely with large and complex documents. (The file for this book was corrupted twice as I wrote it). Keep multiple back-ups of your work on an external hard drive, or online.

6.4 WRITE SIMPLE, CLEAR, CONCISE ENGLISH

Write simply, clearly, and concisely, without pretension. Never distract your reader with your style. When you read something that is well-written, ask yourself why and learn from it. Refer to style manuals for rules. It's okay to break rules, but only if you know what they are.

Use a single space between sentences. Your word processor will do the rest. Check that your grammar, spelling, and punctuation are correct. Learn how to use commas, apostrophes, semicolons, and colons correctly. A good grammar and spell check programme is very useful. Check any instructions you have for whether to use US or British English and be consistent. Many problems arise when translating from other languages (Box 6.2) or applying advice for non-scientific writing to scientific writing.

Make sure that the order and flow of your ideas is logical and maintain the same order throughout your document. For example, if you set out three aims in a report, organise the data analysis section, the results, and the discussion in the same way. Similarly, refer to predictions in the same order throughout the text.

Use lists for related ideas. Use an introductory sentence to establish the context and provide a smooth transition into the first point. Be consistent in lists: *First ... second ... third* or *Firstly ... secondly ... thirdly* are both fine, but don't mix them. Don't start with *first*, then stop, or start at *second*. Check your lists are grammatically correct. For example, *monkey behaviours include eating, sleeping, and moving* works, but *monkey behaviours include eating, sleeping, and to move* does not. Either make your list a single paragraph (if it is short) or give each element its own paragraph. Don't mix the two approaches.

Make every word informative (Box 6.3). Use standard words with established meanings. Rephrase negative statements to be positive (e.g. *Write positively* not *Don't be negative*). Check for common errors and correct them (Box 6.4).

6.5 ENGAGE AND GUIDE YOUR READER

Use the first sentence of a report, section, or chapter to engage your readers. For example, begin with a strong statement about the context of your study, then justify the statement in the opening paragraph. Avoid beginning with *In this chapter ...* or with a review of the previous section, as this can be rather tedious.

Tell your readers what to expect by providing brief summaries of the sequence of topics you will address (**signposts**). For example, if several paragraphs each address one aspect of a topic, preview the list before going into the details.

Walk your reader through your argument so that you convey your message clearly. Don't expect them to figure something out for themselves.

In long documents headings to guide your readers. Ensure that headings are clear, consistent, and logical. You do not need a subheading for each paragraph. Too many headings indicate a lack of logical connection between sections, fragment the text, and interrupt flow.

Box 6.3: Filler words and phrases

Verbose text is frustrating to read and confuses readers. Avoid unnecessary filler words and phrases that you can cut or shorten without loss of information. For example, you don't need to tell your readers that *In this section, I will introduce my research question . . .* or equivalents that are obvious from the structure of the document. This is fine as a reminder to yourself, while you write, but make sure you delete it later.

Use active verbs, not nouns created from verbs or adjectives (**nominalisations**). Nominalisations often end in *-ation* or *-ing* and lead to unnecessarily wordy phrases. For example, use *elevated testosterone* not *an elevation of testosterone*, *is debated* not *is a topic of debate* and *reflects* not *is reflective of*.

Adverbs provide extra information about verbs or adjectives. They can be useful but are often unnecessary. For example, use *remove*, not *remove completely*.

There is no need to describe published studies as *earlier, prior,* or *previous*. They couldn't be anything else. The exception to this is where you contrast earlier and more recent work.

Avoid **pleonasms** – phrases in which one or more words are redundant. For example, *three species* not *three different species, group* not *social group*,[1] *small* not *small in size, red* not *red in colour*, and not *and in addition*.

Trim redundant word pairs. For example, reduce *first and foremost* to *first* and *each and every* to *each* or *every*.

Cut all these:

---

[1] With thanks to Eckhard Heymann.

Box 6.3: (cont.)

As yet
Importantly
In fact
In the process of
It goes without saying
It is important to emphasise/highlight/note that
Studies have shown that
Which is, that were, who are, etc.

Replace this	With this
a number of	*several*
a large number of	*many*
are in agreement with	*agree with*
are known to be	*are*
at this point in time	*now*
as well as	*and*
by means of	*by*
composition of the group	*group composition*
despite the fact that	*although*
due to the fact that	*because*
during the course of	*during*
for the purpose of	*for*
for the reason that	*because*
has been found to be	*is*
have been found to be	*are*
in addition, we	*we also*
in order to	*to*
in proximity to	*near*
is indicative of	*indicates*
it is ... that ...	delete these words, keep the rest, and the phrase will still make sense
not only ... but also ...	*... and ...*
on a daily basis	*daily*
on the other hand	*in contrast*
prior to/previous to	*before*
the fur of the monkey	*the monkey's fur*
was located in	*was in*
there are ... that ...	delete these words, keep the rest, and the phrase will still make sense

Box 6.3: (cont.)

(cont.)

Replace this	With this
the reason for this is the fact that	*this is because*
we are writing	*we write*
we conducted an investigation into (see nominalisations, above)	*we investigated*
we performed an analysis on the data (see nominalisations, above)	*we analysed the data*
we studied a total of 10 mandrills	*we studied 10 mandrills*
we were able to write	*we wrote*
what is clear is that	*it is clear that* (or just cut the phrase)
we conducted a/this study on	*we studied*

FURTHER READING

Trenga B. 2006. *The Curious Case of the Misplaced Modifier: How to Solve the Mysteries of Weak Writing*. Cincinnati, OH: Writer's Digest Books. A concise and simple guide to writing well, including how to identify and avoid the passive voice, vague phrasing, and sentences that are too long.

There's plenty of similar advice on the Internet. Make your own list and share it with colleagues. Challenge yourself to write as concisely as possible.

Box 6.4: Be a pedant

Here's a list of words and phrases that are commonly misused or that are confused with each other:

Accuracy/precision	*Accuracy* is the closeness of a measured value to the true value. *Precision* is the degree to which repeated measurements under unchanged conditions show the same results to each other and is related to reproducibility and repeatability. Accuracy and precision are independent of one another. You can be precise but inaccurate. You can also be accurate but imprecise. You can also be both, or neither.
Aim to/aim at	Use *aim to*.

Box 6.4: (cont.)

Affect/effect	*Affect* is usually used as a verb meaning to influence something. *Effect* is usually used as a noun and means an outcome.
Among/between	Use *among* where the specific items are not named, and *between* for any number of specific (named) items.
And/or	Use *or* not *and/or*, because *or* includes the possibility of both.
Assumption	Often used to refer to a *hypothesis*. Use *hypothesis* in such cases.
As compared with	Use *than* (in comparisons, e.g. *males are larger than females,* not *males are larger as compared with females*).
Attitudes	*Attitudes* are always *to* or *towards* something; they are not *of* something.
Believed	*Believed* is a subjective assessment; replace it with a scientifically supported statement.
Both ... share	*Both ... share* is redundant. Use either *both* or *share*, but don't use both in the same sentence.
But	*But* is a conjunction which expresses contrast, so don't use it at the start of a sentence. Strict grammarians feel the same way about *However*, but I don't.
Change/difference	*Change* means to become different. Authors often refer to *change* when they report a *difference* between two groups. This is not a change; it is a *difference*.
Circa	Use *circa* where exact dates are unknown. In other contexts, use *about* or *approximately* if you cannot be exact.
Cf.	The abbreviation *cf.* means compare, not see. Get the punctuation right.
Compare with/compare to	*Compare with* means to examine differences and similarities; *compare to* means to identify similarities. You probably mean *compare with*.

Box 6.4: (cont.)

Confirm/support	Our data may *support* a prediction; they do not *confirm* it.
Continual/continuous	*Continual* means repeated at intervals. *Continuous* means without interruption.
Correlated with/correlated to	Two variables are correlated *with* one another, not *to* one another.
Data	*Data* are plural; the singular is *datum*.
Decreased, increased	these, and similar comparisons, compare measures made on the same group; they do not compare across groups.
Different from/different than	Use *different from* not *different than*.
Et al./and colleagues	*Et al.* is an abbreviation for *et alia*, which means *and others*. Use *et al.* (get the punctuation right) in parentheses. Use *and colleagues* in the main text.
Fewer/less	*Fewer* means a smaller number of; use it for countable objects (e.g. *fewer monkeys*). *Less* means a smaller amount of; use it for something that can't be counted or doesn't have a plural (e.g. *less fur*) and before adjectives and adverbs.
Focals	We record behaviour in *focal samples*, not *focals*.
Former/latter	Don't use these words, because they require the reader to stop and go back in the text to make sense of what you mean.
Frequency	*Frequency* is the number of times something occurs during a particular period of time (e.g. per min, per hour). It is meaningless without the time units. It is usually synonymous with *rate* in primatology.
Gender/sex	*Genders* are socially constructed cultural categories. *Sex* is biological. They are not synonyms. Use *sex* for animals.[2]
Impact	*Impact* implies force, so don't use it when you mean *effect* or *influence*.

---

[2] Of course, it's more complicated than that.

Box 6.4: (cont.)

Imply/infer	*Imply* means to suggest something indirectly. *Infer* means to deduce or conclude information from evidence and reasoning, rather than from explicit statement of that information.
Important	Never claim that something is *important* without explaining why.
Insignificant	Often used incorrectly in place of *non-significant*.
Mass/weight	*Mass* is usually more appropriate than *weight*, which is mass times acceleration due to gravity.
Methods/methodology	*Methodology* is the theoretical study of *methods*. It is not a synonym for *methods*. Use *methods*.
Mortality/death	*Mortality* is the rate of *death*. If an animal died, that is a death. If two animals died in a year, the *mortality* (not the *mortality rate*) was two per year.
Oestrus/oestrous (UK English) estrus/estrous (US English)	*Oestrus* is the phase of the female *oestrous* cycle when a female is sexually receptive. Females of species with menstrual cycles can be sexually active at any time in their cycle, so the term *estrous* is inappropriate.
Perceptions	*Perceptions* are always *of* something, not *to* something.
Predict/expect	Use *predict* and *prediction*, not *expect* and *expectation*.
Principle/principal	A *principle* is an adjective meaning a general rule; *principal* is an adjective meaning main or most important.
Prefer	*Preference* (or choice) means to like something more than something else. In primatology, we use it when an animal selects something more than expected based on availability (e.g. preference for a fruit; preference for a social partner). It is commonly misused to describe the pattern observed without respect to availability (e.g. dietary budget, distribution of interactions).
Prosimian	Don't use *prosimian*. Use *strepsirrhine* or *tarsier*, as appropriate (see Chapter 7).

Box 6.4: (cont.)

Ranges	The dash in *x–y* means *between x and y*, so you don't need to precede a range with *between* or *from*. Use either *x–y* or *from x to y*.
Rate	*Rate* is the number of times something happens per unit time (e.g. per min, per hour). It is meaningless without the time units. It is usually synonymous with *frequency* in primatology.
Respectively	*Respectively* means *in the given order*. It requires the reader to stop and go back in the text to make sense of what you mean, so don't use it.
Review/revise	Reviewers *review* manuscripts; authors *revise* them. Often confused by non-native English speakers.
Significant	Reserve this for statistical significance, to avoid ambiguity.
Similar to/similar as	Use *similar to*.
Some/certain	Use *some* not *certain*, which suggests a mystery.
Support/accept/prove	Findings *support* a hypothesis or a prediction or fail to do so. We cannot *accept* or *prove* a hypothesis or a prediction.
Study of/study on	We conduct a *study of* a question or species, not a *study on* it.
Such as ... among others/etc.	These are redundant when used in the same list. Do not use them together. The same applies to *e.g., ... etc.* in the same list, including ... *among others*, and *e.g., ... among others*.
Test/confirm/prove/verify	We *test* predictions, we do not *confirm, verify*, or *prove* them.
Therefore/thus	*Therefore* introduces a conclusion that follows what you have just written. *Thus* means *in this way*. The difference is subtle and usually overlooked.
Units	Use units correctly. For example, $20 \times 50 \text{ m}^2$ or $20 \text{ m} \times 50 \text{ m}$ but not $20 \times 50 \text{ m}$.
Use/utilise	Use *use* (why use three syllables when one will do?).
Utilisation	Use *use* (this cuts four syllables).

Box 6.4: (cont.)

Values	*Values concern* something. Don't use *values to* or *values of.*
Varying/various	*Varying* means changing. *Various* means different.
Viz.	*Viz.* is outdated. Use *namely*, if you need it at all.
Which/that	*Which* is for non-restrictive clauses (*The monkey, which I observed, is a pygmy marmoset*). *That* is for restrictive clauses (*The monkey that I observed is a pygmy marmoset*). I find this one difficult.

More Pedantry

There is no need to modify an absolute (e.g. *unique*, not *very unique*).

Make the antecedents of pronouns (e.g. *it, this, their*) clear. For example, *their* in *Vocalisations of mandrills and their use in surveys* has two possible antecedents (vocalisations and mandrills) and suggests you might use mandrills in surveys, although using vocalisations would be more appropriate. Antecedents at the beginning of a sentence are often unclear. Fix this problem by inserting a noun (e.g. I inserted *problem* in this sentence).

Rephrase double negatives as a positive.

Avoid exaggerated language. Be precise and use words you can quantify. Don't use *incredibly, imperative, myriad, vast, vital*, and so on.

Avoid unnecessary qualification. For example, *one possibility is that you may use* . . . includes two qualifiers (*possibility* and *may*) so rephrase as either *one possibility is that* . . . or *you may use* . . .

Complete comparisons with *than* . . . For example, *mandrills are better than baboons*, not just *mandrills are better*.

Compare like with like. For example, *Male canines are larger than females* suggests that male canines are very big indeed, whereas *Male canines are larger than those of females* suggests a more reasonable level of sexual dimorphism.

We describe primates as *folivorous* but we don't describe their diets as *folivorous* because *folivorous* means *leaf-eating*. The same applies to frugivory and other diets.[3]

Be careful when stringing nouns together or using more than one word to modify a noun. Compound modifiers can arise from an

---

[3] With thanks to Jess Rothman.

Box 6.4: (cont.)

attempt at brevity, but often introduce ambiguity. For example, use *the probability of infant survival* not *infant survival probability*. This results in a longer sentence, but one that is easier to read. Clarity is more important than brevity. Use hyphens to prevent confusion. For example, a *permanent staff meeting* is ambiguous. It might be a meeting for permanent staff, but it might also be a permanent meeting for staff.[4]

Check for redundant words (e.g., *both were the same*).

Be consistent with tenses. Use the present tense for published work. Use the past tense to describe your methods, results, and any unpublished data.

FURTHER READING

Clymo RS. 2014. *Reporting Research: A Biologist's Guide to Articles, Talks, and Posters*. Cambridge: Cambridge University Press. Provides a longer list of misused words.

You'll find further lists of common errors in all books on writing. Also see any good dictionary, grammar blogs, and style guides.

6.6 WRITE SENTENCES AND PARAGRAPHS

Write clearly, simply, and precisely using short sentences. Give each sentence a clear subject and a strong, precise verb. Keep the verb (the action) and the subject close together. Stick to one idea per sentence. You can often split a long sentence at *and* or *but*. Don't embed sentences within sentences using parentheses or between dashes. If you need the information, move it into the main text. If you don't need the information in parentheses, cut it.

Combine some short sentences to improve flow. Variation in sentence length aids readability, but long scrambled sentences interfere with effective communication. Deciphering text requires mental energy and tires your reader.

Address a single topic in each paragraph. Begin with a topic sentence that encapsulates the main point of a paragraph, telling the

---

[4] With thanks to Ian Rickard for the example.

reader what to expect. A good way to be sure your topic sentences make sense is to copy and paste them into a new document and check that the topics are clear and that the flow of ideas is logical. Use the rest of the paragraph to expand on your point. End with a conclusion – what you want the reader to remember – but beware of repeating the topic sentence or the content of a paragraph.

Address one idea at a time. Don't interrupt an idea to move on to another one, then return to the first idea.

Use line spacing or indentation to separate paragraphs clearly. Avoid orphan sentences (i.e. a single sentence as a paragraph). Either elaborate on the point or put the sentence into another paragraph. Limit paragraphs to less than two-thirds of a page when double-spaced. Long paragraphs often begin with one topic, then continue into another. Split these into separate paragraphs.

6.7 USE EFFECTIVE TRANSITIONS

Don't rely on headings to make transitions for you. Ensure that the text makes sense without them.

Use transition words to connect phrases and sentences sparingly and accurately. For example, *However, …* indicates that you are about to present a contradiction. You don't need a transition word at the beginning of every sentence. Transition words don't make sense at the beginning of a paragraph.

Use summary nouns to clarify what you're referring to in a new sentence. For example, *This point is important* not *This is important*.

Beginning a new paragraph with a sentence that links it to the last paragraph facilitates flow. You may have been taught to use the last sentence of one paragraph to preview the next, but this doesn't work very well because it changes the topic of the paragraph in the last sentence.

6.8 USE THE ACTIVE VOICE

Use the active voice, not the passive. In other words, use *I ate the donuts* not *The donuts were eaten*. This tells the reader who did the eating.

You may have learned to use the passive voice when writing science, but it is awkward and wordy, and defeats the goal of simplicity. The active voice is shorter, simpler, and less pompous than the passive voice. Active phrases are easier to read, and science has an international audience. To check for the passive voice, try adding *by mandrills* at the end of a phrase. If that makes sense, you need to rephrase with an explicit subject to the verb.

6.9 AVOID REPETITION, REDUNDANCY, AND VERBOSITY

Scientific writing is often needlessly wordy. Use short words where you have a choice, use adjectives and adverbs sparingly, and cut unnecessary words and phrases (Box 6.3). Complex ideas do not need to be presented in a complex way. Use readability tests, but remember that these are not specifically designed for science writing. Avoid including words in an attempt to impress (you won't), don't puzzle your readers, and don't ramble.

Check for redundancy. Statements like *as mentioned above, aforementioned*, and *see below* are either unnecessary or indicate a structural problem. Avoid the need for these phrases by moving all the material on a topic to one place. If you must refer readers elsewhere, refer to a specific section of the text so they don't need to search all the preceding or subsequent text to find the material you're referring to.

Put clauses in the simplest order. Reorder to reduce the number of commas (assuming you have the commas in the right places). For example, you can rephrase *In 1996, I started working in Gabon* to *I started working in Gabon in 1996* to lose a comma. Begin sentences with the main point, not with a qualification. In other words, avoid beginning a sentence with *Although* or *Because*.

Eliminate excessive detail and to make your point, tell readers only what they need to know. For example, you don't need to state the aim of a study you're reviewing, just the relevant conclusion.

Check that your text flows well and never surprise your readers.

6.10 USE JARGON SPARINGLY

Science writing is more formal than writing for a general audience (Box 6.1), and more formal than the language I use in this book. Avoid colloquial phrases (i.e. informal language) and exaggeration (e.g. *vast*). Don't use contractions (e.g. write *don't* as *do not*, unlike in this book).

Be consistent with terms and avoid synonyms. You may have been taught to avoid using the same word twice in a paragraph, but you should not do so in science. If you do, your readers will think they have missed a subtle distinction between the terms.

Use simple words and short phrases rather than obscure words and long phrases. A limited amount of jargon aids communication among experts, but don't use jargon in a misguided attempt to impress.

6.11 USE WORDS CAREFULLY

Check your language for bias and unintentional prejudice and avoid anthropomorphic and value-laden terms (Box 6.5).

6.12 USE ABBREVIATIONS SPARINGLY

Abbreviations include contractions, acronyms, and initials. Use as few abbreviations as possible, because they interfere with the readers' experience. Avoid new or specialist abbreviations that your readers will need to learn or look up when abbreviations recur. These are tempting if you have a word limit, but they reduce readability.

At the beginning of a thesis or dissertation, provide a list of the abbreviations you do use. In journal articles, define all abbreviations at first mention in the abstract and in the main text by giving the full term, then the abbreviation in parentheses. Use only the abbreviation from then on.

Don't abbreviate common names. Instead, give the full common name the first time you mention a species, then define a shortened version of the name for subsequent use. For example, in an article about Western lowland gorillas, use *gorillas*, not *WLG*. Similarly, use *macaque* for *lion-tailed macaque*, not *LTM*; and *tamarin* for *golden-headed lion tamarin* not *GHLT*. Use *primate*, not *NHP*. There is no need to specify that you mean one of the >99% of primate species that are not human.

Use short, explanatory terms that are helpful to your readers, rather than arbitrary terms (e.g. *wet season* rather than *period 1*).

6.13 AVOID LOGICAL FALLACIES

Flawed reasoning undermines your argument. Look out for logical fallacies in your own writing and when you read articles.

Straw man arguments misrepresent or exaggerate an argument to refute it more easily. This can be a deliberate attempt to make a point seem stronger or more novel than it is or can arise from insufficient

Box 6.5: Language matters

The language we use both reflects and influences our thinking. We cannot be completely objective and avoid sociocultural influences on our terminology, but we can reflect on the implications of the language we use, and choose words, metaphors, and analogies carefully. Problematic and persistent biases in primatology include inappropriate anthropomorphism, androcentrism, biased language when discussing human–other primate interactions, conflation of sex and gender, and unintentional prejudice when describing people. This includes widely used terminology. Look for these when reading the literature and avoid them when writing.

Anthropomorphism
We can't avoid using human terms for primate behaviour, because they are the only terms we have. We naturally attribute human characteristics to animals (**anthropomorphism**) and this provides a rich source of hypotheses. However, human terms can have inappropriate connotations and bias our thinking. Researchers who study animal behaviour avoid attributing emotions to animals, and use descriptors with neutral value, but anthropomorphic terms slip in elsewhere in our descriptions. For example, if primates eat crops, they are *crop-feeding* or *crop-foraging*. This is often described as *crop-raiding*, but *raid* is inappropriately anthropomorphic and implies the motivational state of the animals, casting them as thieves or criminals and people as victims.

Primates live in *groups*, not *troops* (troops are military sub-units). Describe primate groups using the number of males and females that compose them. Do not describe a group as a *family*, because this does not help to understand the social organisation or relationships among group members. *Family* describes human social groups of variable compositions, who may or may not live together, and may or may not be biological kin, making it meaningless as a description of an animal group.

Where both males and females mate with multiple members of the opposite sex, describe this as *polygynandry*, not *promiscuity*. *Promiscuity* does not differentiate between male and female behaviour, making it less precise and less useful than polygynandry. Moreover, *promiscuity* implies indiscriminate mating, which is unlikely to be the case, and has gendered implications in everyday parlance.

Box 6.5: (cont.)

Androcentrism
Traditional descriptions of primate groups centre the male point of view (they are **androcentric**). Avoid this by using *one-male, multi-female* to describe a group with this composition, not *harem*, which is androcentric and implies male control of females, as well as anthropomorphic. Similarly, use *one-male, multi-female unit* or *reproductive unit*, not *one-male unit*, because a one-male unit either describes a single animal or ignores the females.

Don't project gendered stereotypes onto primates. For example, female reproductive strategies are not merely *counter-strategies* to male *strategies*.

Assumptions about Interactions between Humans and Animals
Use neutral language to describe interactions between humans and animals. In addition to avoiding the term *crop-raiding*, use *human–animal relations*, not *human–wildlife conflict*, which assumes that relations among species are negative, and precludes positive or neutral relations.

Sex and Gender
Use *sex*, not *gender*, when referring to male and female primates. Sex refers to physical characteristics, while genders are socially constructed roles and identities. Of course, the biological and the sociocultural interact in complex ways, but stick with *sex* for primates.

Writing about People
Check for bias and unintentional prejudice when writing about people. Use gender-neutral terms: *humans* or *people*, not *man*. Refer to *local people* not *locals* or (worse) *natives*. Refer to *ethnic groups*, not *tribes*, and use the names that people use to describe themselves, not those imposed by others. Use *belief*, not *superstition*.

FURTHER READING

Fedigan LM. 1982. *Primate Paradigms: Sex Roles and Social Bonds*. St. Albans, VT: Eden Press. Includes extensive discussion of value-laden and androcentric terminology.
Fedigan LM. 2001. The paradox of feminist primatology: The goddess's discipline? In: Creager ANH, Lunbeck E, Schiebinger L (eds.). *Feminism*

Box 6.5: (cont.)

in Twentieth Century Science, Technology and Medicine. Chicago, IL: The University of Chicago Press. pp. 46–72. A reflection on androcentrism and gender bias in primatology.

Fedigan LM, Jack KM. 2013. Sexual conflict in white-faced capuchins: It's not whether you win or lose. In: Fisher ML, Garcia JR, Chang RS (eds.). *Evolution's Empress: Darwinian Perspectives on Women.* New York: Oxford University Press. pp. 281–303. A review of sexual conflict in capuchins that avoids assumptions about which sex has an advantage.

Hill CM. 2017. Primate crop feeding behavior, crop protection, and conservation. *International Journal of Primatology* 38: 385–400. https://doi.org/10 .1007/s10764–017-9951-3. Includes the need to use neutral language when describing primate crop-feeding.

Karlsson Green K, Madjidian JA. 2011. Active males, reactive females: Stereotypic sex roles in sexual conflict research? *Animal Behaviour* 81: 901–907. https://doi.org/10.1016/j.anbehav.2011.01.033. Shows the prevalence of stereotypes in the sexual conflict literature.

Peterson MN, Birckhead JL, Leong K, Peterson MJ, Peterson TR. 2010. Rearticulating the myth of human–wildlife conflict. *Conservation Letters* 3: 74–82. doi:10.1111/j.1755-263X.2010.00099.x. Argues that the phrase *human–wildlife conflict* is detrimental to human–wildlife coexistence.

understanding of the literature and reliance on secondary material. Check for unsupported generalisations and phrases like *It was previously assumed that* . . . .

Ad hominem arguments attack the author personally, rather than addressing the argument itself. This may be an attempt to discredit an argument by suggesting that the person is not competent, or to question the motives of the author.

Hasty generalisations jump to conclusions based on limited information. Make sure conclusions are based on good evidence.

Circular reasoning (or **begging the question**) occurs when the conclusion is the same as one of the propositions that supports the conclusion. In other words, evidence given in support of a claim assumes that the claim is true, rather than the evidence supporting the claim. In simple forms, this is easy to spot (e.g. A is true because A is true, or A is true because B is true and B is true because A is true), but it can be hidden in a more complex argument, or where arguments are not stated explicitly.

False dichotomies reduce an argument to two possibilities when there are other alternatives, or the two possibilities are not mutually exclusive. For example, proximate and ultimate explanations for the same phenomenon are not alternatives (Section 1.1).

A **non sequitur** is an argument that does not follow from the statements preceding it. This can happen when you don't explain your reasoning clearly.

An **argument from authority** uses the reputation of the author as evidence in favour of a conclusion, rather than the quality of the argument itself.

Questionable cause involves concluding that if two events occur together, then one caused another (Section 5.3). It also includes the **fallacy of the single cause**, which involves assuming that an outcome has a single cause.

The **bandwagon fallacy** involves following majority opinion, for example, choosing a method because other people use it, rather than because it is valid.

6.14 BE CONSISTENT WITH SPECIES NAMES

Use appropriate taxonomic terms (Box 6.6) and be consistent in your use of scientific and common names (Box 6.7). Don't use scientific and common names interchangeably, because this confuses your readers.

Use initial capitals for International Union for Conservation of Nature threat categories, including Threatened, Endangered, and Critically Endangered.

6.15 USE NUMBERS AND UNITS ACCURATELY

Use words for numbers up to and including nine, unless they have units, in which case use figures. Use figures from 10 onwards. Sentences should always begin with words, so spell out numbers at the start of a sentence or rephrase to avoid beginning with numbers. Use words to clarify two sets of numbers, for example *twenty 10 s intervals*.

Use the Système International (SI) for units, including mm, cm, m, km for length; mg, g, kg for mass; μl, ml, l for volume; s, min, h, d for time. Insert a space between numbers and the unit of measure (6 m, 14 ml) except for symbols for angular degrees (°), minutes ('), and seconds ("), which follow immediately after the number with no intervening space. Do not use. After the unit. Don't add s for values greater than one.

Use . not , for the decimal point in English. Use spaces between groups of three for numbers over 9999. Don't use commas for large

Box 6.6: Problematic taxonomic terms

Many primatologists, and major textbooks, use terminology that is incongruent with modern taxonomy. Don't copy them.

Modern phylogenies are based on **monophyletic** groups (clades), which represent evolutionary history and include all the descendants of a single common ancestor. In contrast, **paraphyletic** groups exclude a subset of the descendants of the common ancestor (i.e. they exclude one or more subsidiary clades).

The most common problem is the term **prosimian**. Prosimian includes all the strepsirrhines, and the tarsiers, which are haplorrhines, but excludes all the other haplorrhines, so it is a paraphyletic group. Use strepsirrhine or tarsier, as appropriate. The term **anthropoid**, often used to contrast with prosimian, is okay, because it includes all the descendants of a single common ancestor.

The term **monkeys** contains two monophyletic groups: the New World monkeys and the Old World monkeys. However, it excludes the apes, which descended from the same common ancestor as the monkeys, so it is a paraphyletic group. Resolve this by either using the full terms New World monkeys and Old World monkeys, the terms catarrhine and platyrrhine, or by including apes among the monkeys (not commonly accepted, although accurate).

Both apes and the great apes are paraphyletic if they exclude humans. This is easily resolved by including humans among the great apes.

Also check for common names that reflect an outdated phylogeny. For example, Barbary macaques may be given the common name Barbary apes because they lack a tail. However, they are not apes, so use Barbary macaque. Doucs (*Pygathrix*) are often called douc langurs, but are more closely related to the proboscis monkey (*Nasalis*) and the snub-nosed monkeys (*Rhinopithecus*) than to any of the langurs, making langur inappropriate. Geladas are not baboons, so don't use gelada baboons.

Finally, avoid the term *lesser apes* for gibbons and siamangs. Use *small apes* or *hylobatids*. There is nothing lesser about them.

numbers, to avoid confusion with the decimal point. Write 0.x, not .x, for clarity.

Be consistent with the number of decimal places you use. Avoid **false precision**, when numbers suggest better precision than is

Box 6.7: Taxonomy and conventions for species names

Science gives each species a unique name composed of two words
(e.g. *Homo sapiens*). These are the genus (*Homo*) and the species
(*sapiens*). This scientific name is always written in italics and the
first letter of the genus is always capitalised. The first letter of the
species name is never capitalised. You can use the genus name
alone, but you cannot use the species name alone.

Species are the basic unit of biological classification. Below the
species are sub-species, also italicised and in lower case (e.g. *Homo
sapiens sapiens*). Above the species is a hierarchy of ranks, which are
not written in italics, but are capitalised. These include the
subfamily, family, superfamily, infraorder, parvorder, suborder,
and order. For example: we are the species *Homo sapiens*, family
Hominidae, superfamily Hominoidea, parvorder Catarrhini,
infraorder Simiiformes, suborder Haplorhini, order Primates.
Families are the major ranks within an order.

Vernacular versions of systematic categories are not capitalised.
For example, we (humans) are hominids (from Hominidae),
hominoids (Hominoidea), catarrhines (from Catarrhini),
simians (Simiiformes), haplorrhines (Haplorhini), and primates
(Primates).

In a scientific report, give both the scientific name and the
common name the first time you refer to a species in the text, for
example: mandrills (*Mandrillus sphinx*). Then use the common name
only (mandrills). Alternatively, use the scientific name throughout
the text. There is usually no need to give the authority and date of
publication of the name.

Don't use italics for common names. Only begin common names
with capital letters if they are proper nouns. For example, Grauer's
gorilla is correct, but mountain gorilla is not, and Madame Berthe's
mouse lemur is correct but pygmy mouse lemur is not.

Abbreviating genus names (e.g. *H. sapiens*) often results in
ambiguity. Don't do this unless you have already written out the
name in the same paragraph and check carefully for any
uncertainty.

One unidentified species is sp., more than one is spp. These are
not italicised.

Don't use short names, such as chimps for chimpanzees, orangs
for orangutans, ringtails for ring-tailed lemurs, in scientific writing.

appropriate. For example, a mean can't be more precise than the original measurements are, and a percentage can't be more precise than the data. Include the numbers when presenting percentages. Avoid using percentages when the sample size is small, to avoid giving a false impression of accuracy.

There are many types of average (e.g. mean, median, mode), so always specify which one you report. Always report a measure of the spread around an average and specify the measure you use (Box 5.1).

Use the 24 hour clock for times to avoid ambiguity (i.e. 09:00 h not 9 am). Use words for months (e.g. 01 September 1970) to avoid confusion between different date formats.

6.16 CITE THE LITERATURE AND OTHER SOURCES OF INFORMATION APPROPRIATELY

Include all the literature you cite in your text in a literature cited or references section. Don't include sources you don't cite.

Different journals use different citation and referencing formats. Use a reference manager to format citations and references. Set it up correctly and check the output carefully.

We cite the literature for two reasons: (i) to establish the origin of ideas and credit the original author and (ii) to support our statements with evidence from previous studies.

Acknowledging other people's contributions to the field is an important ethical principle. Cite the original source of an idea or a finding, not a secondary source that refers to it, and not where you last wrote about it yourself. If you can't find the original source (and you have tried), cite the paper in which you read about the work as Author(s) (year), cited in Other Authors (year). This indicates to your readers where you obtained the information and allows them to trace any misunderstandings.

Read the sources you cite and check that the authors report what you state they report. Don't copy the citations another author uses to support a statement without reading them and without checking that they are the most appropriate articles to cite. Copying citations from other articles propagates misinterpretations, simplifications, and errors, and can lead to some studies being unfairly ignored.

Prioritise information, not the author, by citing authors in parentheses at the end of a phrase. In other words, don't open a phrase with *Author (2017)* . . . .

Only use a direct quotation if the words themselves are essential. This is rarely the case. Otherwise, communicate the ideas in your own words. Include the page number if you do use a direct quotation from the literature. Use quotation marks '' for a short quotation and use a new line and indent a longer quote to separate it from the rest of the text.

If a manuscript has been accepted for publication, but is not yet published, cite it as *in press* and include the name of the journal. If a manuscript is under consideration by a journal, cite it as *submitted* and don't give the name of the journal it is submitted to.

Cite published information if it exists, so that the reader can verify that the source you cite supports your claims. If no such information exists, but unpublished data supports your claim, cite this as *unpublished data*, if you have permission from the researchers to do so. Include the names of the researchers in the citation (e.g. Lemmers, Dirks & Setchell, unpublished data).

If someone provides you with information in a conversation or personal email that is not available in the published record, cite this as a **personal communication** with the initials and family name of the communicator and the date (e.g. personal communication JM Setchell, 17 April 2018). Get permission before citing someone as a personal communication. Some journals require this.

If you observed something yourself that is not already described in the literature, cite this as **personal observation**. If there are multiple authors, include who made the observations (e.g. J.M. Setchell, personal observation). If the phenomenon has already been described, cite the literature, not your own observations.

Don't cite unpublished data, personal communications, and personal observations in the reference list, because readers cannot refer to them.

6.17 CHAPTER SUMMARY

In this chapter, we've seen that:

- Writing is an essential component of research and takes practice
- Good writing conveys information to readers as clearly and simply as possible.
- Poor writing obscures meaning and frustrates the reader.
- Reading helps to identify and absorb good practice.

- No one uses scientific English as their first language, but it is far more difficult for non-native speakers of English.
- We should start to write early during a project and expect to revise documents many times.
- We should plan time to obtain feedback and revise our text.
- Every word should be informative and appropriate.
- We should engage and guide readers through the text.
- We should use clear sentences and carefully constructed paragraphs.
- We must choose words carefully and consider our reader.
- We should avoid logical fallacies.
- We cite the literature to acknowledge other researchers' contributions and to support our statements.

Students often come to primatology with a specific theoretical question, or with detailed knowledge of one or a few species they are particularly interested in but lack a general understanding of primates. The next chapter provides that understanding by introducing the order Primates.

6.18 FURTHER READING

Almossawi A. 2014. *An Illustrated Book of Bad Arguments*. Victoria, Australia: Scribe Publications. A short illustrated introduction to logical reasoning and common errors.

Clymo RS. 2014. *Reporting Research: A Biologist's Guide to Articles, Talks, and Posters*. Cambridge: Cambridge University Press. Includes a lot of useful advice on communicating clearly with your audience, with a rather old-fashioned, historical approach.

Greenberg SA. 2009. How citation distortions create unfounded authority: analysis of a citation network. *British Medical Journal* 339: b2680. https://doi.org/10.1136/bmj.b2680. Examines the influence of citation distortions on science.

Hailman JP, Strier KB. 2006. *Planning, Proposing, and Presenting Science Effectively*. Cambridge: Cambridge University Press. Appendix A covers how to write clearly.

Matthews JR, Matthews RW. 2014. *Successful Scientific Writing: A Step-By-Step Guide for the Biological and Medical Sciences*. 4th edn. Cambridge: Cambridge University Press. Includes exercises to improve your writing. Enlivened with amusing quotations and cartoons.

Orwell, G. 1946. *Politics and the English Language*. www.orwell.ru/library/essays/politics/english/e_polit [Accessed 3 January 2019]. Essay including six rules for writing.

Simkin MV, Roychowdhury VP. 2003. Read before you cite! *Complex Systems* 14: 269–274. Estimates that only about 20% of authors who cite a paper have actually read it.

Strunk, W Jr. 1920. *The Elements of Style*. New York: Harcourt, Brace, and Howe. A classic guide to US English. First published privately in 1918. Now in the public domain and freely available on the Internet. Don't follow the advice to use masculine pronouns.

Strunk W Jr, White EB. 1959. *The Elements of Style*. New York: Macmillan. Several subsequent editions. Expanded version of Strunk's original volume.

Trenga B. 2006. *The Curious Case of the Misplaced Modifier: How to Solve the Mysteries of Weak Writing*. Cincinnati, OH: Writer's Digest Books. A concise and simple guide to writing well, including how to identify and avoid the passive voice, vague phrasing, and over-long sentences.

Truss L. 2003. *Eats, Shoots and Leaves: The Zero Tolerance Approach to Punctuation*. London: Profile Books. An entertaining approach to punctuation.

Also see the many academic writing blogs online.

7

Introduction to the Primates

Primates are a group of **eutherian mammals**. (Eutheria are one of the two groups of mammals with living members.) Primates share a set of derived traits that distinguishes them from all other mammals, but these traits are not all unique to primates and no single trait defines the primates. They range in body mass from the 30 g Madame Berthe's mouse lemur (*Microcebus berthae*) to around 250 kg for a male Grauer's gorilla (*Gorilla beringei graueri*). This variation in size is in line with that found in other mammalian orders and is closely associated with what they eat (diet), how they move (locomotion), and how they behave.

In this chapter, I provide a general introduction to the primates and their evolutionary **adaptations** (traits produced by natural selection for their current function), including their distribution and habitats, adaptations to life in the trees, diet and dietary adaptations, brains and sensory traits, life history and reproduction, behaviour, and interactions with other species. I then survey the major groups of primates (Fig. 7.1).

7.1 PRIMATE DISTRIBUTION AND HABITATS

Primates occur naturally on five of the seven continents. Most primates live in mainland Africa, Asia, and South America, and on islands off their coasts. They were formerly widespread in Europe and North America, but now occur only in Central America and south-eastern Mexico. No primates other than humans (*Homo sapiens*) have ever occurred naturally in Australia or Antarctica.

Most primates live in tropical moist forests, occupying strata from the forest floor to the canopy. However, primates also occur in diverse other habitats, including tropical dry forests, montane forests, mangrove vegetation, coniferous forest, broadleaf temperate forests,

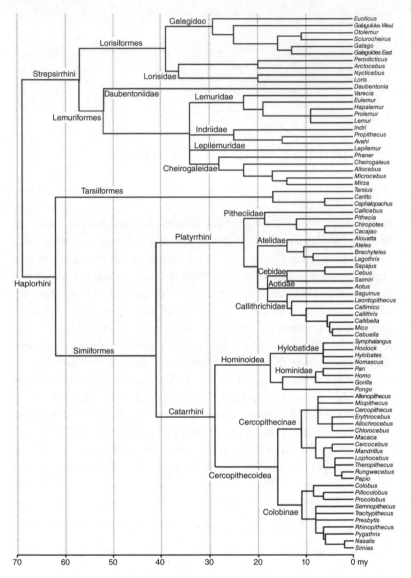

Fig. 7.1 Phylogeny of living primates to genus level, with time in millions of years (my). With thanks to Christian Roos and Dietmar Zinner.

savannahs, grasslands, high-altitude plateaus, and deserts. Some species are partly terrestrial, spending much of their time on the ground, including baboons and ring-tailed lemurs (*Lemur catta*). A few species are fully terrestrial, including geladas (*Theropithecus gelada*).

Primate species distributions vary from extremely limited (e.g. the geographic distribution of the Marohita mouse lemur, *Microcebus marohita*, covers less than 40 km^2 of severely fragmented habitat) to very broad (e.g. rhesus macaques, *Macaca mulatta*, inhabit a large diversity of habitats in Central, South, and Southeast Asia). Humans are globally distributed.

7.2 PRIMATE SKELETONS AND ADAPTATIONS TO LIFE IN THE TREES

The primate skeleton is generalised in comparison with many mammals and retains elements that are reduced or absent in other mammals. These include specialised adaptations for life in the trees. For example, unlike many other mammals, most primates have five digits on each limb (**pentadactyly**) to promote grasping.

Primate hands and feet are further adapted to grasp small curved surfaces (small branches), with flattened nails instead of claws. Highly sensitive pads on the fingertips also enhance coordination and the sense of touch. Some smaller taxa have modified nails that are more like claws, allowing them to use larger tree trunks, including fork-marked lemurs, needle-clawed bushbabies, and callitrichids. Several taxa have specialised grooming claws on some toes (tarsiers, strepsirrhines, and some platyrrhines).

Most primates have opposable thumbs and big toes, which improve stability during arboreal locomotion and allow a precision grip. The thumb is reduced in some species. Some species have grasping (**prehensile**) tails.

Unlike many mammals, primates retain the clavicle (collarbone), permitting shoulder joint mobility, which is important in a three-dimensional arboreal environment.

7.3 PRIMATE DIET AND DIETARY ADAPTATIONS

Primate diets are diverse. They obtain energy by eating plants (**herbivores**) and animals (**carnivores** or **faunivores**). Primate herbivory includes fruit-eating (**frugivores**), leaf-eating (**folivores**), seed-eating (**granivores**), flower-eating (**florivores**), nectar-feeding (**nectarivores**), and tree exudate-eating (**exudivores**). Primate carnivory includes invertebrates (mainly **insectivores**), small vertebrates, larger vertebrates, and other primates (e.g. chimpanzees, *Pan troglyodytes*, hunt and eat various other primate species).

Primates have relatively unspecialised teeth compared to some mammals, reflecting their varied diets. All primates have the same four types of teeth (incisors, canines, premolars, and molars), although the number of each type of tooth (the **dental formula**) varies across taxonomic groups. Adaptations to primate teeth reflect the diet. For example, exudivores have incisors specialised for gnawing holes in trees to access exudates, folivores have shearing crests on the molars to cut leaves into small pieces, frugivores have rounded molar cusps to crush fruits, and insectivores have high points on the molar cusps to puncture the insect exoskeleton.

Primates have a relatively unspecialised alimentary system for mammals, reflecting their generally omnivorous diet. However, folivores have a variety of specialisations for digesting leaves, including large salivary glands and complex guts to promote microbial fermentation.

Primate foods vary in nutritional characteristics and in their distribution and availability across space and time. Primate diet is strongly linked to body size, morphology, and behaviour. Although small animals need less food overall than large animals, they have higher energy requirements relative to their body mass than large animals do. In other words, small animals need more energy per kilogram body mass than large animals do. This means that small animals need a relatively small amount of food, but this must be high-quality food. Insects can provide this and can satisfy the requirements of a small primate. However, large primates cannot catch enough insects to fulfil their needs. Instead, larger primates, which need more food but have relatively lower energy requirements than small primates, forage on leaves. Unlike small primates, whose guts are too short to process plant fibre, large primates can digest leaves. Thus, the smallest primates are exudivores and carnivores. Small- to medium-sized primates are generally insectivore-frugivores, while larger primates are generally folivore-frugivores.

7.4 PRIMATE BRAINS AND SENSORY ADAPTATIONS

Primates have relatively large brains for their body size and the degree of **encephalisation** (the complexity or relative size of the brain) varies greatly across the order. These large brains have been linked to selection for both ecological and social intelligence.

Primates rely extensively on vision and have elaborate visual apparatus in comparison to other mammals. They have forward facing eyes on the front of the skull, unlike most mammals, which have eyes to the side of the skull. This eye position reduces the field of view but results in overlapping visual fields (**binocular vision**), allowing improved distance perception (**stereoscopic vision**). Such three-dimensional perception is an advantage for arboreal animals, allowing them to locate the next branch or food item accurately.

Whereas most eutherian mammals are **dichromatic** and unable to distinguish green and red, primates have a diversity of colour vision abilities, from no colour vision (**monochromacy**) to routine three-colour vision (**trichromacy**).

Primates have a reduced olfactory apparatus relative to other mammals. The snout is reduced, particularly in monkeys and apes. The olfactory region of the brain is also relatively reduced in primates compared to other mammals, although it's not clear whether this is related to olfactory ability. The study of primate olfaction was relatively neglected until quite recently, but we now know that olfaction plays an important role in primate reproduction, communication and food choice.

7.5 PRIMATE LIFE HISTORY AND REPRODUCTION

Like other mammals, primates have a relatively simple **life history** (the series of changes an organism undergoes during its life). Key life stages and events include gestation, lactation, growth and development, reproduction, senescence, and death. Relative investment in growth, reproduction, and maintenance varies across these stages. Compared to other mammals, primates grow and reproduce slowly and have long lives for their body size. Reproductive seasonality is linked to seasonal variation in resource availability. Some species are highly seasonal breeders, some show moderate seasonality, and others reproduce year-round.

Most primates give birth to a single offspring at a time, but some usually bear twins (e.g. most callitrichids) and some can have more offspring in a litter (e.g. dwarf and mouse lemurs, ruffed lemurs). Reproductive investment includes care by mothers, fathers, and other group members (**allocare**). Cooperative breeding, where group members delay their own reproduction and actively help to raise the young of other individuals, has evolved in callitrichids.

We use maturational stages to group individual primates into **age classes** (Box 7.1).

Box 7.1: Getting age classes right

Development is a continuous process, from conception to death. However, we often need to divide the life cycle into phases, to compare like with like. It can be difficult to age wild primates, thanks to their long lifespan and dispersal. So, we define age classes, based on life stages.

Standard age classes with clear definitions are vital for comparison with other studies. However, age class definitions are often not clearly stated, and are not standardised between species, between studies or even between the two sexes in the same study. I give standard definitions here.

Infant

The **infant** stage (infancy) begins at birth, and ends at **weaning**, when an individual no longer depends directly on its mother for survival. Weaning is a process, rather than a sharp transition, as the individual gradually moves from relying on mother's milk to obtaining its own food. Infants are carried and protected by the mother or other caretakers. Their behaviour is dependent on that of the caretaker. Correlates of weaning, and thus the end of infancy, include resumption of cycling by the mother, cessation of suckling, eruption of the first molar, and changes in pelage.

Juvenile

A **juvenile** is a weaned individual who has not yet undergone **puberty**. Puberty is a physiological process leading to the beginning of sexual cycling in females (**menarche**), easily visible in species with sexual swellings, but less so in some other species. In males, puberty is marked by first ejaculation, which is (obviously) difficult to assess, but changes in testicle size are easier to detect. Juveniles are pre-reproductive, obtain their own food, and travel independently. They are often remarkably difficult to identify as individuals.

Adolescent

An **adolescent** is an individual that has entered puberty but is not yet fully grown. Adolescence begins at the onset of puberty. Puberty ends when animals are reproductively mature – in other words when females are capable of bearing an infant to term, and males of siring. However, at this point, they may still be growing.

Adolescence ends when growth is complete. It occurs in both sexes, but researchers often ignore it in females by terming females

Box 7.1: (cont.)

adult once they have borne an infant (i.e. when they are **parous**). However, if these females are still growing, they are adolescent.

Adolescents are also termed **sub-adults** or **young adults**. Some authors group adolescents with fully grown animals, as reproductively mature. However, this groups together animals that are still investing in their own growth with those that are fully grown – two very different life history stages. Some authors lump younger (smaller) adolescents with juveniles, but this is inappropriate, because adolescents have undergone puberty.

Adult

An **adult** is reproductively competent and has ceased to grow (i.e. it is full size). Adults can reproduce, but do not necessarily do so, because reproduction depends on social circumstances in both sexes. For example, reproduction may be physiologically or behaviourally suppressed in subordinate females of cooperatively breeding species and reproductively competent but low-ranking males may not reproduce.

Males and females of some species are radically different in appearance (e.g. mandrills, *Mandrillus sphinx*), while in others they are almost impossible to distinguish (e.g. owl monkeys). Parous females often have elongated teats. In some species, some adult males have two morphs; for example, some adult male orangutans develop large cheek pads and throat sacs (flanged males), while others do not (unflanged males).

We cannot assume that individuals carrying infants are mothers, or that the juveniles follow their mothers when travelling. Other animals also care for young including juveniles, adolescents and adults of both sexes.

As adults age, they gradually deteriorate in physical performance (**senescence**).

FURTHER READING

Setchell JM, Lee PC. 2004. Development and sexual selection in primates. In Kappeler PM, van Schaik C (eds) *Sexual Selection in Primates: New and Comparative Perspectives*. Cambridge, UK: Cambridge University Press. https://doi.org/10.1017/CBO9780511542459.012. Includes definitions of age classes.

7.6 PRIMATE ACTIVITY PATTERNS, LOCOMOTION, AND RANGING PATTERNS

Unlike many other mammals, most primates are active during the day (**diurnal**), although many are active at night (**nocturnal**), and some are active at any time of day or night (**cathemeral**). Diurnal primates tend to rest in the middle of the day.

Primate postures include sitting, standing, lying down, vertical clinging, and hanging by the tail, hand, or foot. Primate locomotion is more diverse than that of any other order of mammals. Primates leap, walk, and run on four limbs on the ground and in trees (**quadrupedalism**), climb, and swing (**brachiate**). Of the extant primates, only humans are habitually bipedal, using only two limbs for walking.

Primate locomotion varies with body size due to the relative size of arboreal substrates and gaps. For example, smaller primates are more likely to leap, and can walk on smaller branches than larger primates, but they have difficulty grasping wide trunks.

The distance a primate moves in 24 h is the **daily path length** or **day range**. The total area used over a longer period (e.g. a year) is the **home range**. The most heavily used area is termed the **core area**. Home ranges are not exclusive and often overlap. **Territories** occur where primates defend their home ranges and there is little or no overlap.

7.7 PRIMATE SOCIAL BEHAVIOUR

Primates are highly social when compared with other animals, with diverse grouping and mating patterns. Descriptions of social systems include the distribution of individuals in time and space (**social organisation**), who mates with whom (**mating system**), and types and patterns of social interactions (**social structure**). These descriptions are often confused in the literature, but are subject to different selection pressures, so it is important to distinguish between them.

7.7.1 Social Organisation

A **social unit** includes animals that share a common range and interact more often with one another than they do with other conspecifics. All primate species interact regularly with their conspecifics (i.e. they are **social**), but the size, composition, and cohesion of the social unit vary greatly. Some species also show a great deal of intra-specific variation in social organisation.

Primate social organisation balances the advantages and disadvantages of being solitary (but not asocial) with those of living in a group in terms of passing on genes to subsequent generations (**reproductive success**). The potential advantages of living in a group include increased protection from predators and improved access to resources including food, mates and offspring care. The potential disadvantages include increased susceptibility to predators, which may be able to detect groups more easily than they can cryptic, solitary animals, increased competition for the same resources (food, mates and offspring care), and increased risk of disease transmission.

Traditional classifications of social organisation focus on behaviour during the active period. Animals may feed and travel alone (e.g. some mouse lemurs, some galagos, some tarsiers, and orangutans), in **pairs** of one adult male and one adult female (e.g. woolly lemurs, titi monkeys, owl monkeys, some callitrichids, and many gibbons and siamangs), in **one-male, multi-female** groups (e.g. most colobine monkeys, most guenons, some howler monkeys, and some gorillas), in **multi-male, one-female** groups (e.g. many callitrichids) or in **multi-male, multi-female** groups (e.g. ring-tailed lemurs, some sifakas, capuchins, squirrel monkeys, woolly monkeys, macaques, most baboons, vervet monkeys, and some colobine monkeys). The number of adult males and adult females found together are key to these descriptions. Juveniles and infants contribute to the overall group size. You may see groups described only by the number of males (e.g. *one-male unit* or *multimale group*), but this is androcentric and fails to describe the group adequately. For example, a one-male unit would be a single animal, and a multi-male group does not distinguish between a group of only males, or a group of males and females (Box 6.5).

These classifications of social organisation work well for most diurnal primates, which tend to sleep in the same groups that they forage in during their active period (although some groups of some species aggregate at night, e.g. proboscis monkeys, *Nasalis larvatus*). However, many primate species are nocturnal, and their social organisation has been neglected, leaving large gaps in our understanding of primate social organisation.

Many nocturnal primates that feed and travel alone share home ranges with conspecifics and many sleep in groups. This led researchers to categorise them as **solitary foragers** – species that are solitary during their nocturnal active period but share sleeping sites during the day. However, this term conceals a great diversity of complex social behaviour, some of which is easily missed by human observers. For example, a primatologist radio-tracking a single animal in the dark may be unaware of other animals around them. Nevertheless, detailed field studies have revealed that animals that feed and travel alone maintain acoustic and olfactory contact with one another and interact with conspecifics regularly. In other words, they are social. Many of the social organisations we see in group-living species also exist in dispersed species. In other words, they may live as dispersed pairs (e.g. some sportive lemurs) or dispersed multi-male, multi-female units (e.g. mouse lemurs) and so on.

In some primate species, groups vary in individual membership over time, as smaller units fission and fuse (e.g. ruffed lemurs, spider monkeys, chimpanzees, bonobos). In other words, groups vary in **fission–fusion dynamics**. You may read descriptions of *fission–fusion societies*, but this description of social organisation does not include information about group composition. Describing the degree of fission–fusion separately has the advantage of describing flexibility in social organisation more broadly, enabling comparative studies across species.

Large multi-male, multi-female groups may also be made up of smaller groups in **multi-level** or **hierarchical societies** (e.g. hamadryas baboons (*Papio hamadryas*), geladas, snub-nosed monkeys, proboscis monkeys). Unlike in societies with high fission–fusion dynamics, these smaller groups have a consistent composition, often with one male and multiple females. They may be nested – with the smallest units aggregating into clans, which then aggregate into bands, which also include males that are not attached to smaller units.

Male group membership varies across the year in some species. Males that are not associated with a bisexual group either live alone, associate with groups of other species, or live in **all-male groups**.

Social organisation also includes whether members of one or both sexes remain in the natal group (**philopatry**), and whether members of one or both sexes disperse prior to reproducing (**dispersal**). Females (some squirrel monkeys, atelids, red colobus, and chimpanzees), males (e.g. some squirrel monkeys, most cercopithecines) or both sexes (e.g. many lemurs, Thomas' leaf monkeys (*Presbytis thomasi*), gorillas) may disperse. Among group-living species, dispersing females may enter established bisexual groups or form new groups and dispersing males may enter established groups, become solitary or form all-male groups.

We cannot make assumptions about one component of a social organisation based on others. For example, hamadryas baboons and geladas both live in multi-male, multi-female groups composed of smaller one-male, multi-female subgroups, but hamadryas baboons show male philopatry while geladas show female philopatry. Thus, it is important to be specific when describing the social organisation of a given group or population.

Use neutral terms when describing primate social organisation to avoid attributing inappropriate meaning. For example, use *one-male, multi-female* to describe a group with this composition, not *harem* (Box 6.5).

7.7.2 Mating Systems

Mating systems describe who mates with whom (the **social mating system**) and the reproductive outcome (the **genetic mating system**). Understanding the social mating system requires detailed behavioural observations while understanding the genetic mating system requires genetic determination of parentage, which is not available for many species.

Like social organisation, mating systems are highly variable both within and across primate species. Descriptions often concentrate on the male's behaviour but describing both male and female behaviour clarifies our thinking and avoids inappropriate focus on only one sex.

Both males and females may mate with one or more members of the opposite sex. Males may mate with one female (**monogyny**) or more than one (**polygyny**). Females may mate with one male (**monandry**) or more than one (**polyandry**). Primates also engage in sexual behaviour with members of the same sex.

In a **monogamous** mating system, both males and females mate with only one member of the opposite sex (i.e. monogyny and monandry, e.g. titi monkeys, owl monkeys, some callitrichids, some gibbons). In a **polyandrous** mating system, one female mates with several males (polyandry) and each male mates only with that one female (monogyny) (e.g. some marmosets and tamarins). In a **polygynous** mating system, one male mates with multiple females (polygyny), each of whom mates only with that male (monandry) (e.g. some howler monkeys, many colobines, gorillas). In **polygynandry**, both males and females mate with multiple partners (i.e. polygyny and polyandry, e.g. some mouse lemurs, some atelids, many papionins, chimpanzees). Avoid the value-laden term *promiscuous* for primates (Box 6.5). Bear in mind the temporal component to descriptions of mating systems; conventionally they apply within a breeding season, but not all primates breed seasonally.

In some species, sexual selection leads to **sexual dimorphism** with males larger than females in body size and canine size, and sex differences in **pelage** (fur) colour and length and exaggerated ornaments (e.g. skin colour). Sperm competition leads to the evolution of large testes in species where multiple males mate with the same female during a single receptive period.

7.7.3 Social Structure

We can understand social structure at three levels. At the first level, individual pairs of animals (**dyads**) interact with one another. These interactions can involve spatial proximity, affiliation, and **agonism** (aggression and submission). At the second level, the content, quality, and patterns of these social interactions constitute **relationships**. At the third level, the content, quality, and patterns of relationships constitute the social structure. Social network analysis is a popular method for describing social structures, with individual animals as nodes in a network and ties connecting them to one another.

Relationships can be relationships between females, between males, and between the sexes. They can be among kin or non-kin. Both sexes compete with individuals of the same sex for mates, and with all conspecifics for food. Both sexes may also cooperate to gain and main-tain access to resources. Relationships vary in strength and tolerance. Some species have clear, linear dominance hierarchies, revealed by patterns of submission, in one or both sexes (e.g. ring-tailed lemurs, some capuchins, female callitrichids, cercopithecines). Other more egalitarian species rarely show agonistic behaviour and hierarchies are difficult or impossible to discern (e.g. red-fronted lemurs, *Eulemur rufi-frons*, some squirrel monkeys, muriquis, female blue monkeys, *Cerco-pithecus mitis*). In some species, female hierarchies are based on kinship (e.g. cercopithecines); in others they are not (e.g. colobines). Females dominate males in many lemur species.

In group-living species, relationships between groups range from lengthy peaceful interactions (e.g. bonobos) to lethal aggression (e.g. male chimpanzees).

7.7.4 Socioecological Models

Socioecological models focus on relationships between ecology (the distribution of food) and grouping patterns. These models predict relationships among females based on the degree to which females can monopolise resources within and between groups, influ-encing philopatry, preferential treatment of kin (**nepotism**), and the nature of social relationships. These relationships include relationships between females, relationships between males based on the distribution of females, and relationships between the sexes based on sexual selection and sexual conflict, with potentially complex feedback loops where male strategies and female strategies interact.

7.8 PRIMATE INTERACTIONS WITH OTHER SPECIES

Primates are members of **ecological communities**, coexisting and interacting with other species. Interactions between species in a community can be beneficial to both organisms (**mutualism**), beneficial to one and neutral for the other (**commensalism**), beneficial to one and negative for the other (**predation** and **parasitism**), neutral for one and negative for the other (**amensalism**), or negative for both (**competition**).

Primates interact with other animals as competitors, prey and predators. Primates compete for resources with members of their own species and with other species occurring in the same area (**sympatric species**). Species coexist by playing different ecological roles (occupying different **ecological niches**). Some primate species form mixed-species groups (**polyspecific associations**), which may bring predator protection or foraging benefits. Other mammals and birds also appear to benefit from association with primates. Primates are prey for a wide range of predators, including large cats, fossa (*Cryptoprocta ferox*), domestic dogs (*Canis lupus familiaris*), snakes, monitor lizards, raptors, and humans. Carnivorous primates prey on invertebrates and vertebrates.

Primates form mutualistic relationships with plants, dispersing seeds and pollinating flowers. However, they also prey on leaves, seeds, and flowers, with a negative effect on plant survival and reproduction.

Like all animals and plants, primates are host to, and have coevolved with, microbial organisms including bacteria, archaea, protists, fungi, and viruses. These organisms may be beneficial mutualists (e.g. gut microbiota play important roles in digestion and the development of the immune system), non-harmful commensals, or parasites (where a microbe benefits from the interaction but harms the host). Some parasites have only a minor effect on the host, while others cause serious disease or debilitation. **Ectoparasites** live on the surface of the host, attached to the skin or fur, or burrow into subcutaneous flesh. **Endoparasites** live inside the host, in the gastrointestinal tract, blood, and other tissues.

7.9 PRIMATE DIVERSITY AND TAXONOMY

Primates are currently thought to have diverged from other mammals approximately sixty-five million years ago, during the Palaeocene, making them a very recent group of organisms adapted to fill different ecological niches (**adaptive radiation**). Primates are one of the most species-rich groups of mammals, surpassed only by bats (Chiroptera) and rodents (Rodentia). In August 2018, the International Union for Conservation of Nature recognised 511 primate species in 79 genera, and 702 species and sub-species.

Taxonomy is the system of names which we use to communicate about specific groups of organisms. It describes the organisms we study and the relationships between them. Taxonomy is a science, and species numbers, names and the relationships among them change as evidence accrues and is assessed. In other words, phylogenies are hypotheses. Taxonomists use traits present in organisms, but absent in their ancestors (**derived traits**) to reconstruct evolutionary history (**phylogeny**). They debate species concepts, revisions and the validity of newly described species. The number of primate species recognised, and their names, change as newly discovered species are described, and currently accepted species are re-examined in the light of new data. There is no *right* answer, there is only our current understanding, in which the relationships between the higher order taxonomic groups are well-established, but those among some families are not yet clear (e.g. among some Lemuriformes and Platyrrhini, Fig. 7.1).

The number of primate species recognised has increased markedly since the early 1990s, largely due to taxonomic revisions based on the **Phylogenetic Species Concept**. In contrast to the **Biological Species Concept**, which defines species based on reproductive isolation, the Phylogenetic Species Concept defines species as the smallest possible diagnosable units. In other words, species have features that can be uniquely recognised and cannot be divided into smaller diagnosable subunits. Adoption of the Phylogenetic Species Concept has led to splitting of widespread species into previously unrecognised species with much more restricted distributions, the redefinition of sub-species as to species, and the splitting of genera. Such splitting is controversial, but the species described are testable hypotheses.

When writing, support statements on primate taxonomy with references to the primary, peer-reviewed, scientific literature.

Primates are divided into two suborders: the **strepsirrhines**[1] (Strepsirrhini, wet-nosed primates) and the **haplorrhines** (Haplorrhini, dry-nosed primates).

7.9.1 Strepsirrhines

The strepsirrhines are sometimes called *lower* or *basal* primates, but this terminology reflects the misconception that they are primitive in relation to other *higher* primates and hinders our understanding of their evolutionary adaptations, so we shouldn't use it (Box 6.6). Strepsirrhines retain the wet nose and the reflective layer behind the eye that reflects light and enhances night vision (**tapetum lucidum**) found in other mammals. They have a bar of bone around the eye socket (a postorbital bar), which is a derived primate trait, but do not have a complete bony eye socket. Except for the aye-aye (*Daubentonia madagascariensis*), strepsirrhines have specialised dental structures to facilitate grooming (toothcombs). They also have grooming claws. Strepsirrhines show a great diversity of visual abilities. Monochromacy, dichromacy, and polymorphic colour vision have all been described in multiple species. Scent-marking is important in their communication.

Historically, the social behaviour of nocturnal strepsirrhines was greatly underestimated. However, the types of social systems among solitary foragers may be as diverse as those among diurnal primates that forage in groups.

The strepsirrhines include the **Lorisiformes** and the **Lemuriformes**. The Lorisiformes include the galagos (Galagidae) of sub-Saharan Africa and the lorises, pottos, and angwantibos (Lorisidae) of Africa, India, and Southeast Asia. Lorisiformes are all small and nocturnal. Males are usually slightly larger than females. Galagos are specialised leapers, while lorises, pottos, and angwantibos are adapted for slow climbing. Lorisiformes eat insects, gum, and fruit. Some lorises are highly faunivorous. Some lorisids produce a toxin which females apply to their infant's fur as protection, and which gives them a toxic bite. Much remains to be discovered about lorisiform social systems, but existing studies suggest that they are solitary foragers with overlapping home ranges and sleep in groups. Some galagos spend some of their active period in groups and show a high frequency of social interaction.

---

[1] The number of *r*s in strepsirrhine is debated. I follow Groves in the *International Encyclopedia of Primatology* (pp. 1185–1186), in using a double *r* in Strepsirrhini/strepsirrhine and Haplorrhini/haplorrhine.

Lorisiform mating systems are largely unknown, but females mate with multiple males in some species.

The Lemuriformes (the lemurs of Madagascar) are socially and ecologically diverse, having evolved on Madagascar in the absence of competition for ecological niches with the haplorrhines. Living (extant) lemurs are small to medium in size, varying in body mass from 30 g (Madame Berthe's mouse lemur) to 7 kg (indri, *Indri indri*). Some recently extinct species, now only known from partially fossilised (**sub-fossil**) remains, reached more than 150 kg. In other words, until recently, lemurs covered the entire size range of the primates. Lemurs are sexually monomorphic or females are slightly larger than males. They can be nocturnal, diurnal, or cathemeral. They are mainly arboreal, although ring-tailed lemurs are semi-terrestrial and some recently extinct species were probably terrestrial.

The Lemuriformes include five families: the Daubentoniidae, with just one extant species, the aye-aye; Lemuridae; Indriidae; Lepilemuridae; and Cheirogaleidae. Some species are able to reduce their metabolic rate and often their body temperature (**heterothermy**), for less than 24 h (**daily torpor**), several days (**prolonged torpor**), or longer (**hibernation**). Lemurs have diverse diets, including fruit, leaves and animal prey, and a diversity of locomotor behaviour, including leaping and quadrupedalism.

Lemurs have a wide variety of social systems, ranging from solitary foragers who share sleeping sites (e.g. some mouse lemurs), to multi-male, multi-female groups (e.g. ring-tailed lemurs). Social organisation varies between ecologically similar and closely related species (e.g. among mouse lemurs) and some species show great intra-specific variation in social organisation (e.g. Verreaux's sifaka, *Propithecus verreauxi*). Some species show high fission–fusion dynamics (e.g. ruffed lemurs), but there are no reports of multi-level social organisation in lemurs.

Females are dominant over males in many species of lemur. Lemurs are highly seasonal breeders and their mating systems include polygyny, polgynandry, and monogamy. Some species show marked male reproductive skew despite a lack of sexual dimorphism in body mass and canine size (e.g. red-fronted lemurs, Verreaux's sifaka), possibly because sexual selection selects for other traits in these arboreal animals, such as locomotor performance.

7.9.2 Haplorrhines

Haplorrhines have larger brain sizes relative to body size than strepsirrhines and reduced olfactory apparatus and behaviour. They have a fully

enclosed eye socket and lack the tapetum lucidum found in strepsirrhines. The haplorrhines include the tarsiers (**Tarsiiformes**) of Southeast Asia, and the **Simiiformes**.

Tarsiers are small, arboreal, highly adapted for leaping, and sexually monomorphic. They have enormous eyes and monochrome vision
and are primarily insectivorous and nocturnal. They have a single
infant, which is very large relative to their body size, which the mother
parks by itself while she forages. Group composition can vary within a
species. For example, Gursky's spectral tarsiers (*Tarsius spectrumgurskyae*)
sleep in small groups of a bisexual adult pair and immatures, or sometimes one adult male, two adult females and immatures. Although
tarsiers spend much of their nocturnal activity period alone, they also
spend a substantial proportion of it associated with their group mates.
Tarsiers are often placed in the suborder prosimians with the strepsirrhines, based on perceived behavioural similarities, but this terminology is misleading as it obscures the tarsier's independent evolution,
so we shouldn't use it (Box 6.6).

The Simiiformes are split into the **platyrrhines** or **New World
monkeys** (Platyrrhini, flat-nosed primates) of south and meso-America
and the **catarrhines** (Catarrhini, down-nosed primates) of Africa and
Southeast Asia.

Platyrrhines are small to medium sized, with sexual
dimorphism ranging from males that are much larger than females,
through sexual monomorphism to females slightly larger than males.
They are arboreal and inhabit a wide range of forest habitats. They
include five families: Pitheciidae, Atelidae, Cebidae, Aotidae and Cal-
litrichidae. Atelidae and some Cebidae have prehensile tails. Platyrrhines have a diversity of colour vision abilities. Most species show
polymorphic trichromacy in which males and some females are
dichromats and other females are trichomats, but owl monkeys are
monochromats and howler monkeys are trichromats. Many platyrrhines have odour-producing skin glands and show conspicuous scent-
marking behaviours. Platyrrhine diets are variable, including exudates, fruit, leaves, invertebrates, and vertebrates. Only the owl
monkeys are nocturnal or cathemeral; all other species are diurnal.
Platyrrhine social organisation ranges from pairs (e.g. owl monkeys)
to large multi-male, multi-female groups with high fission–fusion
dynamics (e.g. spider monkeys). There are no solitary platyrrhines.
Platyrrhines show the full range of primate mating systems, including
monogamy, polyandry, polygyny, and polygynandry. Some species
show extensive paternal care of offspring and cooperative breeding

(titi monkeys, owl monkeys, callitrichines), which are not found in other primate radiations.

Catarrhines are all diurnal, medium to large primates and are often sexually dimorphic with males larger than females. They show less olfactory specialism than platyrrhines and strepsirrhines and have trichromatic vision. Most species live in groups. Catarrhines include the **apes** (Hominoidea) and the **Old World monkeys** (Cercopithecoidea). Apes have highly mobile shoulder joints, permitting brachiation, and large brains relative to their body size. Apes include the gibbons or small apes (Hylobatidae) of Asia and the great apes (Hominidae).

Gibbons are relatively small-bodied, arboreal, and frugivorous. They have specialised adaptations for brachiation, including very long arms and hook-like hands. Gibbons live in small groups, often of one male, one female and their offspring, and defend territories with vocal displays. They are sexually monomorphic. Gibbons are often termed *lesser* apes, but this erroneously implies that they are less important than other apes, so use *small* (Box 6.6).

Great apes occur in Africa and Asia and include the globally distributed humans. They are all large and sexually dimorphic, with males larger than females, although to different degrees. Great apes vary in arboreality and terrestriality, but all except humans spend substantial amounts of time in trees. Great apes walk on their fists (orangutans), knuckles (gorillas, chimpanzees, and bonobos), or on the soles of their feet (humans). Great apes are frugivorous, folivorous, and carnivorous. Their social systems include solitary individuals (orangutans), multi-female groups with one or more adult males (gorillas) and multi-male, multi-female groups with high fission–fusion dynamics (chimpanzees and bonobos).

Finally, Old World monkeys live in Asia and Africa, and are diverse in habitat, diet, and social behaviour. They are medium to large in size and often highly sexually dimorphic with males larger than females. They can be arboreal or terrestrial, or both. Old World monkey groups include multiple females and one or more males. Some species form multi-level societies. Females often remain with their mothers for life, forming groups of females related through the maternal line (**matrilines**). Old World monkeys are further split into cercopithecines (Cercopithecinae), with simple digestive systems and cheek pouches to store food and digest starch, and colobines (Colobinae), with specialised dentition and complex stomachs to digest leaves.

7.10 CHAPTER SUMMARY

In this chapter, we've seen that:

- Primates are diverse in body mass, habitat, diet, morphology, ecology, and behaviour.
- Primates are specialised for life in trees, but not all primates live in trees.
- Primates have comparatively large brains for their body size, relative to other mammals, and this varies across the order.
- Primates rely extensively on vision and have a diversity of colour vision abilities.
- The primate olfactory apparatus is reduced in comparison with other mammals and varies across the order. Nevertheless, olfaction is important in primate reproduction, communication, and feeding ecology.
- Primates have relatively slow life histories.
- Descriptions of primate social organisation and mating systems are commonly confused but are subject to different selection pressures, so we should distinguish them carefully.
- Nocturnal primates that feed and travel alone often sleep in groups during the day and have complex social lives, albeit ones that are more difficult for us to study than those of diurnal primates.
- Primate social systems are extremely variable within and across species, and much is still unknown.
- We should use neutral terms when describing primate social behaviour.
- Primate taxonomy is fairly well resolved, but taxonomy is a science, and our understanding improves over time. Some out-of-date and inappropriate terms persist in the literature.

The next chapter reviews the reasons why we study primates.

7.11 FURTHER READING

Agnani P, Kauffman C, Hayes LD, Schradin C. 2018. Intra-specific variation in social organization of Strepsirrhines. *American Journal of Primatology* 80: e22758. https://doi.org/10.1002/ajp.22758. A review of the primary literature shows that many strepsirrhines display complex and often variable social organisations and that only 7% of species are exclusively solitary.
Aureli F, Schaffner CM, Boesch C, Bearder SK, Call J, Chapman CA, Connor R, Di Fiore A, Dunbar RIM, Henzi SP, Holekamp K, Korstjens AH, Layton R, Lee PC,

Lehmann J, Manson JH, Ramos-Fernandez G, Strier KB, Van Schaik CP. 2008. Fission–fusion dynamics: new research frameworks. *Current Anthropology* 49: 627–654. Explains why we should use *fission–fusion dynamics* rather than using the term *fission–fusion* to describe a type of social system.

Boyd R, Silk JB. 2014. *How Humans Evolved*. 7th edn. New York: W. W. Norton & Company. Includes chapters on primate diversity and ecology, primate mating systems, primate life histories and the evolution of intelligence, and primate evolution. Uses mating system terms for social units.

Fleagle JG. 2013. *Primate Adaptation and Evolution*. 3rd edn. New York: Academic Press. Textbook with details of primate adaptation and evolution.*

Fleagle J (ed). 2014. Special Issue on identifying primate species. *Evolutionary Anthropology* 23: 1–40. Twelve essays on how we define and identify primate species.

Fleagle JG, Janson C, Reed K (eds). 1999. *Primate Communities*. Cambridge, UK: Cambridge University Press. Reviews the composition, behaviour, and ecology of primate communities in Africa, Asia, Madagascar, and South America.

Groves C. 2008. *Extended Family: Long Lost Cousins. A Personal Look at the History of Primatology*. Arlington, VA: Conservation International. A history of the knowledge and understanding of non-human primates seen through the eyes of the late Colin Groves.

Groves C. 2012. Species concept in primates. *American Journal of Primatology* 74: 687–691. https://doi.org/10.1002/ajp.22035. A review of species definitions, focussing on primates.

Groves C. 2017. Primates (Taxonomy). In Fuentes A (ed), *The International Encyclopedia of Primatology*. Hoboken, NJ: John Wiley & Sons, pp. 1103–1110. https://doi.org/10.1002/9781119179313.wbprim0045. An explanation of taxonomy, with an annotated taxonomy of primates.

Haraway DJ. 1989. *Primate Visions: Gender, Race, and Nature in the World of Modern Science*. New York: Routledge. A feminist history of primatology in the twentieth century.

Kappeler P, Cuozzo, F, Fitchel C, Ganzhorn J, Gursky-Doyen S, Irwin M, Ichino S, Lawler R, Nekaris K, Ramanamanjato J, Radespiel U, Sauther M, Wright P, Zimmermann E. 2017. Long-term field studies of lemurs, lorises and tarsiers. *Journal of Mammalogy* 98: 661–669. https://doi.org/10.1093/jmammal/gyx013. Reviews the social and ecological diversity of strepsirrhines, with a focus on long-term studies. Highlights the dearth of long-term studies on tarsiers (1), lorises (1), galagos (0), and pottos (0).

Mitani JC, Call J, Kappeler PM, Palombit RA, Silk JB. 2012. *The Evolution of Primate Societies*. Chicago, IL: University of Chicago Press. Synthesises what we know about primate behaviour and socioecology. A hefty and indispensable volume intended as a sequel to the hugely influential 1987 volume *Primate Societies* (Smuts BB, Cheney DL, Seyfarth RM, Wrangham R, Struhsaker TT, Chicago, IL: University of Chicago Press). Part 1 provides very useful, detailed reviews of what we know about each of the major primate taxa. Parts 2–5 address the adaptive problems primates face.

Roos C, Zinner D. 2017. Primate phylogeny. In Fuentes A (ed), *The International Encyclopedia of Primatology*. Hoboken, NJ: John Wiley & Sons, pp. 1063–1067. https://doi.org/10.1002/9781119179313.wbprim0394. An explanation of phylogenetic reconstruction, with a simplified primate phylogeny.

Rowe N, Myers M. 2016. *All the World's Primates*. 2nd edn. Charlestown, RI: Pogonias Press. Wonderful images of the world's primates and a wealth of natural history information. See also the website https://alltheworldsprimates.org.

Strier KB. 1994. Myth of the typical primate. *Yearbook of Physical Anthropology* 37: 233–271. https://doi.org/10.1002/ajpa.1330370609. Reviews the problems associated with generalising from studies of a few species of primate to describe the typical primate.

Strier KB. 2017. *Primate Behavioral Ecology*. 5th edn. New York: Routledge. Primate behaviour and socioecology.*

Tecot SR, Singletary B, Eadie. 2015. Why 'monogamy' isn't good enough. *American Journal of Primatology* 78: 340–354. https://doi.org/10.1002/ajp.22412. A reminder of the need to distinguish between social organisation, mating system, and social structure, focussing on strepsirrhines.

*These books use the prosimian/anthropoid distinction rather than strepsirrhine/haplorrhine. Don't copy this.

8

Why Study Primates?

We may already be convinced of the value of studying primates, but we often need to convince others of that value in proposals, reports, and papers. This chapter covers the reasons to study primates, including appreciation of their fascinating diversity and adaptations, their ecological functions, their evolutionary relationship with humans, their sociocultural importance, concern for their captive welfare and their conservation status.

8.1 PRIMATES ARE INTERESTING IN THEIR OWN RIGHT

Primates are highly diverse (Chapter 7). Field studies of primates can be logistically complex and suffer from small sample sizes, but many creative and tenacious primatologists have overcome these difficulties, making primates one of the best-studied mammalian orders. As a result, primates provide an excellent source of comparative information on diverse, closely related species, with the potential to yield important insights into evolution and adaptation.

In addition to their diversity, primates are highly social, with relatively advanced intelligence. These traits lead to research questions specific to primates, including why some species form groups, the costs and benefits of different types of social bonds, and the selection pressures that underlie intelligence. Primate sociality also complicates the application of theoretical models developed for less social orders to primates. For example, detailed social knowledge of potential mates and rivals in some primate species has implications for sexual selection theory developed for other taxa.

Although most genera of primates have been subject to some study, **taxonomic bias** in the literature skews our understanding of primate adaptations. The literature is highly skewed towards the Old World monkeys, primarily to the genera *Macaca* and *Papio*, reflecting the importance of these genera as biomedical models. Studies are also biased to open country, ground-dwelling, diurnal species, because these are relatively easy to study and are thought to provide a model for human evolution. The African apes are also over-represented, due to their phylogenetic relatedness to humans, although they are a very small proportion of the extant primates. Studies of previously neglected taxa will continue to improve our understanding of primate diversity. Moreover, studies of primates in habitats that have traditionally been neglected, including anthropogenic habitats, allow us to investigate their behavioural plasticity, as well as having important implications for primate conservation (Box 8.1).

Box 8.1: The myth of the pristine

Primatologists studying the ecology and behaviour of living primates have traditionally prioritised study populations living in *intact*, *pristine*, *undisturbed*, or even *virgin* environments, in an effort to understand primates in the environment in which they are thought to have evolved. However, such terms are problematic for several reasons.

First, the animals we study inhabit what remains of their historical range. These are often areas that are less accessible to humans and may also be marginal habitats for the primates. We cannot assume that this is the environment in which they evolved. Study sites also change over time.

Second, we don't choose study groups randomly from a population. Instead, we often choose groups which are already partially or fully habituated to human presence or are easily accessible for other reasons. In other words, we don't choose *undisturbed* animals.

Third, terms like *pristine* reflect an unhelpful dichotomy between humans and the natural world. Humans are part of nature and all environments are affected by humans. In some cases, this **anthropogenic influence** is obvious, such as primates that live near human habitation, foraging on crops and in kitchens. Humans also hunt primates, modify or destroy their habitat, feed them either directly (provisioning) or indirectly (by planting crops), or where primates inhabit urban areas. In other cases, anthropogenic influence is minimal, where primates live in native habitat, forage exclusively on wild foods, are unhabituated, have no interactions with humans (including researchers), have no human predation, co-occur with a full guild of indigenous predators, and have no introduced predators. However, even the most remote places are influenced by human activity, both historically and currently, through the presence of indigenous and other peoples and anthropogenic climate change. Most primate field sites are subject to at least some anthropogenic alteration (we're there, after all). Don't be seduced by the concept of the great explorer venturing into unknown territory. It may be unknown to you, but it is not unknown to humans.

Rather than ranking environments in terms of human modification, we should describe the exact conditions under which study animals live, including anthropogenic effects and historical influences, to inform our understanding of primate behavioural plasticity. Careful description of the conditions our study animals live in are more useful than general terms like *pristine*. Understanding the influence of species on one another, including our own relationships with other species, is an important aspect of both fundamental primatology and conservation.

FURTHER READING

Fedigan LM. 1992. *Primate Paradigms: Sex Roles and Social Bonds*. Chicago: University of Chicago Press. Discusses the problem of defining *natural* behaviour on pages 47–49.

McKinney T. 2015. A classification system for describing anthropogenic influence on nonhuman primate populations. *American Journal of Primatology* 77: 715–726. https://doi.org/10.1002/ajp.22395. A very useful analysis of the different dimensions of anthropogenic influence, and a standardised system for describing conditions at a study site.

Rees A. 2006. A place that answers questions: Primatological field sites and the making of authentic observations. *Studies in History and Philosophy of Biological and Biomedical Sciences* 37: 311–333. https://doi.org/10.1016/j.shpsc.2006.03.008. Examines the idealised notion of the field site as a natural place and how it has shaped primatology.

8.2 PRIMATES HAVE IMPORTANT ECOLOGICAL FUNCTIONS

Many primates play key roles in ecosystems as mutualists, prey, hosts to parasites, predators of plants and animals, and competitors (Section 7.8). They influence plant survival and reproductive success as consumers of plant parts, seed dispersers, pollinators, and insectivores. In particular, seed dispersers help forest to regenerate, and some primates disperse species that are important to human food security.

8.3 PRIMATES HELP US TO UNDERSTAND OUR OWN EVOLUTION

The more than 500 other primate species are our closest living biological relatives. Our common evolutionary history means that we share many characteristics and studying primates can help to understand our own anatomy, physiology, cognition, life history, and behaviour. Studies of extant primates also provide models for understanding the behaviour of our extinct relatives, for whom we have only partial hard tissue remains. We must beware, however, of treating other primates as though they are inferior versions of humans; they are superbly adapted to their own niches (Box 8.2).

Box 8.2: The problem of the great chain of being

The great chain of being, or *scala naturae*, is a view of nature derived from classical Greek philosophy. It suggests that nature forms a hierarchy of advancement and value, from *lower* to *higher* organisms. This concept of a ladder results in the common misconception that some modern animals represent the ancestors of others (e.g. that humans evolved from our closest living relatives, the chimpanzees and bonobos (*Pan paniscus*)). The *scala naturae* also underlies the incorrect description of species that retain primitive (i.e. ancestral) traits as primitive species.

Evolution by natural selection has no aim. Rather than a ladder, all extant branches of the tree of life are equally advanced and of equal rank. The *scala naturae* obscures the exquisite adaptation of each species to its own niche. Humans did not evolve from chimpanzees. Instead, both modern species share a common ancestor, which is now extinct.

The *scala naturae* is reflected in many primate taxonomies, which appear to progress towards the great apes and, ultimately, humans. I deliberately ended the review of primate diversity in Chapter 7 with Old World monkeys. I could have ended with any of the primate taxa.

8.4 PRIMATES ARE IMPORTANT TO HUMAN HEALTH

The same evolutionary relationships that make primates good models for the understanding of human evolution underpin their use in biomedical research. Such research is controversial, but research using primate models has helped researchers to understand many human medical conditions. The study of health and disease in wild primates can shed light on the origins, evolution, and transmission of pathogens, with implications for human health and primate conservation. For example, **zoonoses** are infectious diseases in other animals that can naturally infect humans, and **anthroponoses** are infectious diseases of humans that can infect other animals, such as primates.

8.5 PRIMATES ARE SOCIALLY AND CULTURALLY IMPORTANT

Humans have always co-existed with other primates. Primates play important ecological, social, and cultural roles in many societies. They feature as objects of symbolic thought, metaphors, sacred figures, ancestors, kin, pets, performers, sources of medicine, food, and study subjects. Human relations with primates with whom we share habitat include protection, amused tolerance when primates forage on crops in some cases and retaliation in other cases, predation, and both intentional and incidental provisioning.

Primates contribute to human economies both positively, via primate-related tourism, and negatively, when they damage crops and property. Humans also keep other primates as companions, status symbols, assistants, and zoo exhibits, and use them in biomedical research and entertainment.

These complex interactions have implications for both explorations of interspecies relationships and for primate welfare and conservation.

8.6 PRIMATES NEED SPECIALISED CARE IN CAPTIVITY

Captive primates are housed in zoos, wildlife centres, and sanctuaries, and thousands of animals are used annually in laboratory research.

Ethical responsibilities require us to provide captive animals with environments and diets that promote their optimal welfare. This includes acquisition, transport, housing, husbandry, positive reinforcement training, and colony management. Well-designed scientific research can inform all aspects of care. Appropriately designed research on wild animals can provide information on the diet, nutritional needs, normal behaviour patterns, suitable environments, and physical health and condition.

8.7 PRIMATES ARE THREATENED WITH EXTINCTION

Recent (2017) estimates suggest that 60% of primate species are threatened with extinction and 75% have declining populations. The level of threat varies by region: 87% of species in Madagascar are threatened, 73% in Asia, 37% in mainland Africa, and 36% in the Neotropics. The major threat in all areas is habitat loss, due to habitat conversion for agriculture and cattle ranching, timber extraction, oil and gas drilling, mining, dam building, and road construction. Other threats include hunting of primates for local consumption or for trade. Trade in live primates often involves the death of many others. Primates are also killed in retaliation for foraging on or damaging crops. Disease and climate change add to the list of threats.

Threats to primates threaten biodiversity more generally, due to their ecological functions (Section 8.2). For example, primate extinction may endanger the future of plant species, by depriving them of seed dispersers and pollinators, resulting in changes in forest structure and regeneration. Forests with fewer primates show marked effects on tree species that rely on large frugivores for seed dispersal.

It's easy to despair, faced with this situation, but conservation requires hope. Reversing population declines will require widespread changes in human behaviour. We must reduce demand for the products that drive habitat conversion and promote sustainable practices. We must tackle the inequalities, political instability, organised crime, and corruption that destroy the opportunities of many who live alongside primates. We must work with those who share their land with primates to protect and connect large areas of habitat and promote land-sparing and land-sharing in recognition of the fact that protected areas are not enough to conserve biodiversity. We must enforce laws to eradicate illegal trade and eliminate the demand for wild primates as meat and pets.

8.8 CHAPTER SUMMARY

In this chapter, we've seen that:

- Studies of primates provide important insights into our understanding of evolution and adaptation.
- The literature is skewed towards relatively few primate species, with plenty of possibilities for further study.
- We need to study species in a variety of habitats to understand their plasticity.
- Studies of primates shed light on our own evolution and the behaviour of extinct species.
- Studies of primates help us understand human health.
- Our own species' complex interactions with primates are a fertile ground for investigation.
- We have an ethical responsibility to promote the optimal welfare of captive primates.
- The conservation crisis is serious and requires action.

In the next chapter we look at how we identify a research question.

8.9 FURTHER READING

Boyd R, Silk JB. 2014. *How Humans Evolved*. 7th edn. New York: WW Norton & Company. Addresses how studying primates informs our understanding of human evolution.

Estrada A, Garber PA, Rylands AB, Roos C, Fernandez-Duque E, Di Fiore A, Nekaris KAI, Nijman V, Heymann EW, Lambert JE, Rovero F, Barelli C, Setchell JM, Gillespie TR, Mittermeier RA, Arregoitia LV, de Guinea M, Gouveia S, Dobrovolski R, Shanee S, Shanee N, Boyle SA, Fuentes A, MacKinnon KC, Amato KR, Meyer ALS, Wich S, Sussman RW, Pan R, Kone I, Li B. 2017. Impending extinction crisis of the world's primates: Why primates matter. *Science Advances* 3: e1600946. https://doi.org/10.1126/sciadv.1600946. Reviews the conservation status of primates, concluding that the situation is dire, but there is still hope. Includes why primates matter.

Fuentes A. 2012. Ethnoprimatology and the anthropology of the human–primate interface. *Annual Review of Anthropology* 41: 101–117. https://doi.org/10.1146/annurev-anthro-092611-145808. Reviews the study of human interactions with primates.

Phillips KA, Bales KL, Capitanio JP, Conley A, Czoty PW, Hart BA, Hopkins WD, Hu S-L, Miller LA, Nader MA, Nathanielsz PW, Rogers J, Shively CA, Voytko ML. 2014. Why primate models matter. *American Journal of Primatology* 76:

801–827. https://doi.org/10.1002/ajp.22281. Reviews why primates are used in biomedical research.

Rosenthal MF, Gertler M, Hamilton AD, Prasad S, Andrade MCB. 2017. Taxonomic bias in animal behaviour publications. *Animal Behaviour* 127: 83–89. https://doi.org/10.1016/j.anbehav.2017.02.017. Addresses the existence and implications of taxonomic bias in research.

Wich SA, Marshall AJ (eds.). 2016. *Introduction to Primate Conservation*. Oxford: Oxford University Press. Details of the threats to primates and potential solutions.

9

Identifying a Research Question

A clearly formulated research question is vital in science because it determines the data we need to collect, the methods we use and, ultimately, the success of a project. Developing a research question is an iterative process of reading and thinking, as we define a problem and specify the contribution we hope to make to resolving it. This is not easy, and we learn through experience, and (if we're lucky) from our mentors.

In this chapter, I first describe different types of research, then explain how we use case studies to address broader research questions. Linked to this is the variety of locations in which we can study primates. Next, I explore where questions come from and examine what makes a good research question. Finally, I explain why reading is essential to the development of research ideas.

9.1 FUNDAMENTAL AND APPLIED RESEARCH

Research serves two purposes. **Fundamental research** (**basic** or **pure research**) aims to contribute to our theoretical understanding of how the world works. It is driven by curiosity and generates new ideas. **Applied research** aims to address real-world problems. In primatology, these problems are often welfare or conservation concerns.

9.2 GENERAL RESEARCH QUESTIONS AND SPECIFIC CASE STUDIES

Research questions are theoretical. They address something that we do not yet know. The theoretical question is always broader than the particular case study in which we choose to examine them. In other words, research questions are the big picture.

Case studies may be individual primate species (a **study species**), a community of species (a **study system**), a particular primate radiation, or all the primates. Study species and systems are also described as **model species** or **model systems**. These terms describe a species or system that is particularly well-studied or tractable, and from which we can extrapolate to other species and systems. Some questions are only relevant to some taxa – those that possess the trait of interest. Primatologists interested in human evolution may focus on the species most closely related to us, although a wider comparative approach can be extremely valuable.

9.3 WHERE WE STUDY PRIMATES

We study live animals at field sites, in captive facilities, and in zoos, and obtain data from specimens in museum collections, laboratory analysis, databases, and the literature. Each type of location and approach has advantages and disadvantages and places different constraints on the questions we can test and the conclusions we can draw.

Studies in different settings are often complementary and a combination of approaches tells us more than any single approach can. For example, field observations inspire carefully designed captive experiments, then field experiments translate these experiments to more natural conditions. Studying primates under a variety of conditions tests the validity and generality of our conclusions and improves our understanding of primate behavioural plasticity.

The prevalent *wild* vs. *captive* dichotomy conceals great diversity in living conditions for the primates we study. Studies of live primates can range from the relatively controlled conditions of the laboratory to unhabituated wild primates, via zoo collections, captive colonies, provisioned populations, and habituated animals. We may study free-ranging, enclosed, or caged animals, within or outside their natural range. They may be wild-feeding, provisioned, or a combination of the two. They may live in environments with and without predators, both natural and introduced. Each setting has advantages and disadvantages and the questions we can ask depend on the conditions. Don't fall prey to automatic assumptions that one setting is better than another (Box 8.1).

Our choice of where we study primates depends on the questions we want to address, and on practicality, so we need to bear in mind what is possible as we explore and develop research questions.

9.4 WHERE DO RESEARCH QUESTIONS COME FROM?

Research ideas begin with something we're interested in, which we narrow to a topic, and from there to a question that we can address. They come from theory, our own observations of animals, and a variety of other sources.

If our research is applied, then it is essential to work closely with the people and organisations who are involved in or might be affected by our work and understand their priorities. In other words, we should involve **stakeholders** as partners when planning the research. This will ensure that our work is realistic and relevant to their needs. We cannot assume that people will read reports and act on our recommendations if they were not involved in the project design. Animal care centres, zoos, and sanctuaries may prioritise research that will provide insight into problems they face and will benefit their animals.

9.4.1 Beginning with Theory

Some primatologists are fascinated by a theoretical concept and choose a suitable study taxon in which to study it. In this case, the choice of case study is based on a combination of suitability for the question and practicality. The availability and choice of case study then leads us to revise our question until we arrive at a practical combination that allows us to answer a question. This approach can allow us to ask sophisticated questions about species that are relatively easy to study and for which we already have a great deal of relevant background knowledge. However, such an approach can also lead us to focus on a few tractable species that may not represent other primates, skewing our understanding of the underlying theoretical question. Such bias can also have practical implications for welfare and conservation, if we apply knowledge derived from one species to another without first establishing whether such generalisation is appropriate.

We might aim to test an entirely new idea, although this is relatively rare. More commonly, we test an idea from the literature that has not yet been tested. Replicating and extending previous findings are essential to scientific progress (Section 1.1) and tests of the generality of established paradigms provide fertile ground for new research (Box 1.2). For example, we might test theory developed for mammals, birds, or fish on primates, or test whether theory developed for Old World monkeys can be generalised to New World monkeys or lemurs, or vice versa.

The limitations and future directions of published studies provide useful research ideas. Perhaps we can use more sophisticated methods or accumulated data to re-examine a question from some years ago. Are the old questions really settled? We might propose a test that is more substantial than previous studies, or more rigorous, with a new study design, improved methods or a much-improved sample size.

We also ask comparative questions, comparing observations across species. This can be based on existing data collated from the literature, newly collected data, or both. Where we collect new data, careful choice of species for data collection (**phylogenetic targeting**) improves the study.

If we begin with theory, it is crucial to develop a deep understanding of the study species to ask appropriate questions, design an appropriate study, and interpret our findings.

9.4.2 Beginning with Our Own Observations

Observing primates is a great source of inspiration. It is essential to know our study species well to understand whether a theoretical question is appropriate or relevant. Formulating good research ideas based on observations may seem serendipitous, but chance favours the prepared mind. Good observers constantly ask themselves questions about what they observe, and whether it fits with current theory. Observations also tell us what is practical and relevant to our study species.

Some primatologists begin with a deep desire to understand a particular species. For example, we might be specifically interested in primates in a particular area. We might also be interested in a species with unusual traits or one that is understudied or endangered, or we may be concerned about a species' captive welfare. In such cases, we might begin with a specific observation of a particular species and use this to pose a more general theoretical question. In all cases, it is essential to read and think more broadly than the particular species and to understand what we know about the research question in general.

There may be good reasons why a species is not already well-studied. It may be challenging to study and we may not be able to ask sophisticated theoretical questions, at least not initially. Nevertheless, all long-term studies begin as short-term studies, and studies of new species are important to test the extent to which theoretical models developed using other species apply more generally. Moreover, technological advances allow us to ask previously intractable questions.

If we choose our study system first, our choice constrains the research questions we can address. If a chosen species does not lend itself to addressing the question we're most interested in, we must consider changing to a different study species. Similarly, we may need to change study site.

Over time, primatologists often develop deep interest in and familiarity with a study species. This also allows us to reap the benefits of working at an established study site and accumulating long-term data. Nevertheless, it is vital to ask which theoretical questions we can contribute to, not simply what is feasible at our study site. Visiting other study sites to observe other species is extremely useful to stimulate new questions.

9.4.3 Other Sources of Research Questions

In addition to being inspired by a theoretical concept or a study species, we might also be interested in developing a new method or testing an existing method under new circumstances (e.g. moving laboratory analysis to the field).

Further sources of research questions are the research needs and priorities identified in conservation action plans and management plans. In such cases, it is essential to develop the project with relevant stakeholders and decision-makers to maximise its effectiveness and contribute to evidence-based decision-making.

Funding bodies may also state priority topics. For example, national research bodies often have strategic priorities, and conservation funds might target the least-known or most threatened species.

Supervisors (advisors) may suggest a question, or we might be employed to develop a specific project. In such cases, the initial project description may be as simple as a few sentences or a paragraph or may include a detailed proposal and work plan.

Finally, questions can also come from discussion with other researchers. For example, a question at a conference, or a discussion in a journal club or in an online forum can direct you to new areas of enquiry.

9.5 GOOD RESEARCH QUESTIONS

There are no right or wrong research questions. There are, however, substantial and trivial questions. A substantial question contributes to our theoretical understanding of primates. It has broad relevance, beyond the case study we choose to investigate, and answering it leads to further questions. A trivial question has a simple factual answer that we cannot generalise and leads to no further questions.

An original (or novel) study is one that has not been done before. At one level, all studies are original, because the study has not been done before at that time and place. However, a more useful criterion is whether a study adds significantly to what we knew before.

We should be able to explain the significance of our question. In other words, we should be able to explain what makes the question compelling and why it should interest the academic community and the broader public. How will it enrich our understanding of primates? How does the study integrate with what we already know? What will the implications of our findings be? Who should care?

We should also be able to explain why our study taxon provides a new, and informative, case study that will improve our understanding of a general question in primatology or another field. Unless the data are directly relevant to conservation, the fact that we don't know anything about a question in our particular study taxon is not usually a good justification for a study. Instead, we need to explain what it is about that taxon that makes it a particularly interesting case study in a broader context.

Describing and identifying patterns are crucial steps in posing a question, but good research questions seek an explanation for those patterns. In other words, they ask how or why something happens, in addition to describing it.

A good question is also tractable. In other words, we must be able to answer it with the time and resources available to us. There may be good reasons why we don't already know the answer to a question. Maybe it's difficult to address. If so, look for methodological developments that will help you to overcome these difficulties. If there are none, you need to rethink.

Finally, it is vital to study a topic we are genuinely interested in, because data collection can be tedious, analysis challenging, and writing frustrating, so we need strong motivation.

9.6 DEVELOPING A RESEARCH QUESTION

To develop and refine a research question, we need to obtain a broad theoretical understanding of the topic we're interested in, discover what scientists already know, and identify what remains unknown. We need to know which hypotheses have been tested and falsified, which are supported and how well they are supported. We need to determine which studies need replicating (Box 1.2), which need to be extended, and which assumptions need to be tested. These knowledge frontiers are where we can contribute new understanding. We do all this via a thorough review of the literature.

Reading also tells us which methods work, although what doesn't work may not get published and talking with other researchers is a better source of information.

As we explore a topic further, we notice new things, and pursue new ideas. Some of these ideas lead nowhere, and that's okay. We all develop and discard many possibilities before choosing which to pursue.

9.7 CHAPTER SUMMARY

In this chapter, we've seen that:

- A clearly formulated research question is vital to the success of a project.
- Research questions are conceptual and are always broader than the particular case study in which we examine them.
- If our questions are applied, we must collaborate with the people we wish to influence from the beginning of the project.
- Research questions arise from reading the literature, our own observations, and a variety of other sources.
- The appropriate case study and location for a project depend on the question.
- Reading more broadly than the particular case study is essential to determine what we already know about a research question and the limitations to that understanding.
- We often explore and abandon several potential questions before choosing one to pursue.

Box 9.1 addresses common problems when developing research questions and how to resolve them.

Box 9.1: Common problems when developing research
questions and how to resolve them

Common problems in developing research questions include the
following issues.

Questions That Are Too Simple

A question is too simple if it has a simple, factual answer which
does not advance our theoretical understanding of primates. Ask
yourself what broader theoretical concept underlies your question.
What would the results of your study mean for primatology (or
another discipline) more generally? If you can't answer this, you
probably need to think again.

Questions That Are Too Ambitious

Some interesting questions cannot be addressed. This might be
because the question is too broad, or because we cannot address it
effectively. Broad questions can be a good place to start, but our
final research questions must be answerable, and we must be able
to address them in the time we have available. We can't do
everything in one study (fortunately, otherwise scientists would
be out of a job).

   If we cannot address a question effectively, we should choose a
different question, not conduct a study that will not address the
question.

Questions That Are Too Vague

Some research questions are too vague to address. You can identify
a vague question by attempting to propose answers to it
(hypotheses). If you can't propose an answer, then you need to
clarify and refine the question. Reading will help to focus and refine
your interest to a question that you can address.

Beginning with the Methods

Students often find it easier to think about what they could
measure than to explore theoretical concepts. When asked about
their research question, they respond by talking about methods
they could use, rather than posing a theoretical question. While
methods and logistics are important, it's essential to begin with a
good understanding of the theory that underpins a question, and to
choose the most appropriate study design and methods, not simply
those we are familiar with.

Box 9.1: (cont.)

Weak Rationales
Researchers often use the number of existing studies to justify the need for a study. For example, they justify a new study by saying that *Few studies have investigated question x* or *Question y is under studied*. However, this is not a good measure of importance. A topic may be understudied because it is of little interest, and an ingenious study of a well-studied topic can improve our understanding markedly. Instead, we need to explain why a question is important and what it contributes to our understanding of theory.

Similarly, if a specific question has not yet been addressed in a particular study species, we need to explain which features of a species make it particularly interesting. It is not enough that no such study yet exists.

Developing the rationale for a study, and defining the contribution it makes, is not always easy but is important.

Starting with the Case Study
Some researchers begin by thinking about a specific study species and posing questions about that species. This is fine as a start, but we must broaden the focus so that we pose a general question, with the species as a specific case study.

The next chapter addresses the question of how to search the literature to find out what scientists already know.

9.8 FURTHER READING

Arnold C, Nunn CL. 2010. Phylogenetic targeting of research effort in evolutionary biology. *The American Naturalist* 176: 601–612. https://doi.org/10.1086/656490. Explains how to select target species for comparative study.

Hailman JP, Strier KB. 2006. *Planning, Proposing, and Presenting Science Effectively*. Cambridge: Cambridge University Press. Chapter 1 covers finding a problem.

Karban R, Huntzonger M, Pearse IS. 2014. *How to Do Ecology: A Concise Handbook*. 2nd edn. Princeton, NJ: Princeton University. Chapter 3 covers picking and developing a research question.

Stamp Dawkins M. 2007. *Observing Animal Behaviour: Design and Analysis of Quantitative Data*. Oxford: Oxford University Press. Chapter 2 covers asking the right question, with a specific focus on observational studies of animal behaviour.

Finding Out What We Know

We find out what scientists know about a topic by searching the scientific literature. Literature searches range from a preliminary search to find out what we know about a general area to a specific search on a precise topic. As we explore a research topic, we focus our searches to identify the main open questions, the hypotheses proposed and the support for them, potential model systems and methods, and the experts in the field. Broad background reading is also fundamental preparation for a study because most studies do not go as planned and we often need to identify new research questions as we progress.

I begin this chapter with sources of information we have available, then describe how we identify search terms and assess the quality of the literature we find. I explain the importance of reading broadly and how to choose what to read, and end with how we keep up with the literature.

10.1 SOURCES OF INFORMATION

Peer-reviewed journal articles are the primary source of scientific information. They contain details of the research and reflect the current state of understanding. Secondary sources like good textbooks and encyclopaedias are a good first start when investigating a topic, but they date quickly and reflect the interests and understanding of the authors. Book chapters are also useful, but peer-reviewed journal articles are standard in our field, so I concentrate on them here.

Articles in academic journals are scrutinised by researchers in the same field before they are published (Box 3.1) to ensure the validity of the findings, the quality of the manuscript, and the fit to the journal's aim and scope. Journals range from generalist publications like *Nature* and *Science*, which are highly selective and aim to publish only the most important findings, to specialised publications that focus on a particular topic (e.g. anatomy, animal cognition, behavioural ecology, evolution, conservation, ecology) or taxon (e.g. primates).

Journal articles include primary research articles, short commentaries, reviews, and book reviews. Of these, primary research articles describe new research on a specific question, with original data and the conclusions of the researchers who conducted it. They include detailed methods and results.

Commentaries on high-profile publications can be useful to understand the strengths and weaknesses of the study.

Reviews provide an overview of a field or topic, summarise the data, synthesise the conclusions of many studies and usually provide more detailed background than primary research articles. Good reviews provide road maps for future research. Check both the primatological literature (*Evolutionary Anthropology* and *Yearbook of Physical Anthropology*, and the major primate journals) and review journals in your field (e.g. the *Trends* journals) for these. Special issues of the major primatology journals can be useful updates on a research topic.

Best practice guidelines produced by committees of experts such as the International Union for the Conservation of Nature are very useful sources of information on sampling design, data collection, and analysis for conservation interventions and monitoring, including reintroduction and translocation.

The **grey literature**, including reports produced by nongovernmental organisations, dissertations, and manuscripts that have not yet been subject to peer review (**pre-prints** re), can be very useful, but is not peer-reviewed and varies in quality.

10.2 IDENTIFYING SEARCH TERMS

Once we have identified a topic of interest, we can formulate a search strategy and search terms. This is a skill and takes practice.

Web of Science is the gold standard database for literature searches but requires a subscription. If you are affiliated with an institution, they may subscribe. Google Scholar is freely accessible in most countries and can give a good overview of the literature available on a topic. Try different tools and see what works for you, but don't rely on just one source. Databases differ, so check exactly how each one works.

Begin with the key terms and phrases related to your topic or research question. It can take time to identify these (Box 10.1). Include synonyms and related terms. Use **truncation** where a word has variable endings (e.g. primate* will find primate, primates), and **wildcards** where spelling varies (e.g. colo?r will retrieve color and colour). Focus on the research topic to find relevant studies in other taxa in addition to primates.

Combine search terms using **Boolean operators** to target your search. AND finds references which contain both terms (e.g. primate* AND conservation). OR finds references which contain either word (e.g. monkey OR lemur). NOT (or a − sign) excludes records with a specific word. Quotation marks return exact phrases.

Once you have some search terms, run a search and evaluate the results. Are they relevant to your search? If not, revise your search terms and try again. If there are too many results, then use more specific search terms. If there are too few results, then broaden your search. Check the results for search terms you have missed and add these to your list.

Searches often link you directly to the article, although the full text may not be available (Box 10.1). Searches also link to the articles that cite a publication. Once you find an article that addresses the topic you're interested in, search forward to articles that cite it, and scrutinise the introduction and literature cited for further relevant articles. Keep an eye on your question as you explore. The literature is endless and it's easy to go off track.

Librarians are expert at literature searches. Online interest groups can be useful sources of advice if you are stuck. You risk annoying people if you ask questions you can answer with a simple Internet search.

10.3 ASSESSING THE QUALITY OF THE LITERATURE YOU FIND

Depending on the database or search engine you use, your literature search may return content ranging from unregulated websites to peer-reviewed journal articles. Concentrate on peer-reviewed work, which has been evaluated by reviewers for validity and quality.

There is no simple way to assess research quality without developing your own critical skills and reading the article. (We'll cover this in Chapter 11.) Peer review improves, but does not guarantee, the quality of a report. The reputation of a journal is a useful indicator of quality. Journal **impact factors** (Box 10.2) are often used to assess the quality of

Box 10.1: Common problems in literature searches and how to resolve them

Common problems in literature searches include the following points:

There's Too Much to Read

If you find too much information, this may be because your search is too general (e.g. primat* AND ecolog*). If so, then go back to your research question and terms. Can you narrow them down? Look at some of the records. Are you picking up other subject areas? If so, refocus your search.

There Are No Other Studies to Read

Congratulations; you've asked an original question. You do need to know the context to that question, though. Are your keywords too specific? Have you, for example, only looked for studies of the question in particular species? Try removing the specifics of your search to find studies of similar questions on other taxa.

I Don't Have Access to the Full Article

Access to journal articles is often restricted to users who have paid a subscription to the publisher. You can access the title, authors, abstract, and keywords, but the full text is unavailable if your library doesn't subscribe to the journal, or you don't have access to a library. You can access the full text in other ways. Researchers often post full-text versions of their articles online on their website, in an institutional depository, or on academic social networks. To find these, you can use a web-browser extension that hunts for papers in repositories, or put the exact article title, in quotation marks, into an Internet search. You can also request articles using social media. (Such peer-to-peer sharing may violate copyright rules.)

Some learned societies provide free or reduced-cost subscriptions to their journals for researchers from low- and middle-income countries. Publishers may also waive or reduce subscription fees for institutions in such countries.

You can also email the **corresponding author** and request a copy of the article. The corresponding author's contact information is listed with the abstract. He or she should be happy that someone is interested in the work, but will be less happy if work has been made freely available online, and you haven't checked before asking for it.

Box 10.1: (cont.)

Here's suggested text for a request:

Subject: Request for copy of article
Dear Dr [insert family name]

> *I am very interested in your article [insert the title] in [insert the journal or book] but I cannot access the full text. I would be very grateful for an electronic copy of this article.*

> > *Thank you in advance for your help*
> > *Your name*
> > *Your status and affiliation*

Use the author's correct title. If you don't know what the title is, use Dr. Dr makes no assumption about gender. If your contact doesn't have a PhD, they'll be flattered. If they are Prof, then they are also Dr.

You could add a brief explanation of why you are interested in the author's work. This is a great way of contacting experts in your field for the first time.

Request specific articles. Don't simply request a copy of everything an author has ever written, or everything they have written on a topic. This suggests that you haven't given much thought to the request.

When the author responds, thank him or her.

---

an individual article, but this is inappropriate because the impact factor relates to the journal, not the individual article. The number of times an individual article is cited in other publications is useful but does not necessarily indicate quality; some articles are unjustifiably neglected, and others are cited because they are flawed.

10.4 READING BROADLY

Students often focus their reading on a particular study species, genus, or geographical area. This is important background to a case study, but it is essential to read more broadly within and beyond the primates. Primatologists have much to learn from those working on similar questions in other organisms. Reading outside primatology is also

Box 10.2: Journal impact factors

The impact factor is the most commonly used of several measures of a journal's influence. Impact factors are often used inappropriately.

How Are Impact Factors Calculated?

Impact factors are calculated annually. The impact factor of a journal in a calendar year is the number of citations in journals indexed by Thomson Scientific made during that year to any article published in that journal during the previous two years, divided by the total number of citable articles published in the journal during those two previous years. Note the discrepancy between the nominator (citations of *any* article) and the denominator (*citable* articles).

Problems with Impact Factors

Impact factors measure the influence of a journal and journals are commonly ranked by their impact factor. However, this is problematic for several reasons. First, impact factors do not take variation in citations across articles into account and can be highly skewed if some articles are highly cited while others are not. Review articles are more likely to be cited than primary research articles. Moreover, citations indicate the popularity of an article, rather than its quality. Highly cited articles may be high quality, but they also include those that are heavily criticised.

Second, impact factors can be manipulated by questionable editorial policies such as requiring that authors cite other articles in the same journal.

Third, impact factors cannot be used to compare journals across fields, or even across sub-fields, because different disciplines have different citation practices and move at different speeds. (Remember that the impact factor is based on only the previous 2 years.)

Fourth, the data used to calculate impact factors are not transparent and are not openly available. Impact factors are not necessarily reproducible.

Impact Factors Do Not Measure the Quality of Individual Articles or Researchers

Impact factors compare journals. They are not designed to evaluate the quality of research in an individual article. They are not a good measure of the quality of an individual article because citation

Box 10.2: (cont.)

counts vary greatly across articles published in a journal, and because impact factors refer only to a short period of the journal's history. Nevertheless, impact factors are very commonly misused by managers and employers seeking simple metrics to evaluate research and researchers. This has a knock-on effect on researchers, who select journals based on impact factor, rather than relevance and the audience they wish to reach. This, in turn, delays publication of reports and overburdens editors and reviewers.

Put simply, it is wrong to assume that articles published in journals with higher impact factors are of higher quality than those published in journals with lower impact factors.

FURTHER READING

Hicks D, Wouters P, Waltman L, de Rijcke S, Rafols I. 2015. The Leiden Manifesto for research metrics. *Nature* 520: 429–431. https://doi.org/10.1038/520429a. Introduces ten principles to guide metrics-based assessment of research.

The San Francisco Declaration on Research Assessment. https://sfdora.org [Accessed 9 January 2019]. A set of recommendations for improving the ways we evaluate scientific research.

Seglen PO. 1997. Why the impact factor of journals should not be used for evaluating research. *British Medical Journal* 314: 498–502. https://doi.org/10.1136/bmj.314.7079.497. An article published more than 20 years ago, explaining why impact factors are not appropriate for evaluating research. If anything, the practice is now more pervasive.

essential where we wish to incorporate elements of other disciplines (e.g. endocrinology, parasitology, botany) into our work because we must understand the science that underlies the methods we employ. Reading outside your sub-discipline is an excellent source of inspiration for creative research approaches and study design.

10.5 CHOOSING WHAT TO READ

You can't read everything, so once you have a list of articles on a topic, think about what you want to know and prioritise the articles directly related to that question. Use the title, abstract, and keywords to assess whether an article is relevant to your interests, but don't use these to evaluate the findings. Skim an article quickly to decide whether to read it in depth.

You need to read both the original work in an area and the most recent studies to gain a full understanding of a question. Reading older articles is important to gain a solid grounding in your discipline, and to avoid proposing ideas that have already been suggested. Don't assume that current studies interpret the original work correctly. Often authors copy references without reading the original, repeating, and magnifying misconceptions.

10.6 KEEPING UP WITH THE LITERATURE

New research is published all the time, so you need to keep up with the current state of knowledge in your area of interest. Find a routine. Subscribe to feeds and electronic tables of contents for journals. Do regular searches with your keywords. Create citation alerts. Social media can alert you to exciting new work, but don't rely only on this to identify new studies. It can take a while for a study to be published, so keep up with conferences, too. Attend, check the programme and abstracts online, or follow the event on social media.

10.7 CHAPTER SUMMARY

In this chapter, we've seen that:

- Reading the literature is key to finding out what scientists already know.
- Literature searches range from a preliminary search on a topic to a specific search on a precise question.
- Textbooks are good general background reading and reviews provide a useful overview of a field, but we must read the primary literature to understand research in detail.
- Identifying appropriate search terms takes practice.
- The reputation of a journal is a useful indicator of quality, but the only way to truly assess the quality of an article is to read it critically.
- We must read broadly, beyond studies of our own study species, and beyond primates, to understand what scientists know about a question, and how to address it.
- It's easy to get lost in the literature, so we must prioritise and focus on articles that are relevant to our aims.

- We must read older articles to avoid proposing ideas that have already been suggested.
- We must read the latest articles to understand the current state of our knowledge on a topic.
- We can keep up with new research by setting up alerts and conducting regular literature searches.

Once we've identified articles that are relevant to our current needs, we need to read them. In the next chapter, we look at how to do this.

10.8 FURTHER READING

International Union for the Conservation of Nature Reintroduction Specialist Group. *iucnsscrsg.org* [Accessed 3 January 2019]. General and taxon-specific guidelines for reintroduction.

Pain E. 2016. How to keep up with the scientific literature. *Science Careers* blog. https://doi.org/10.1126/science.caredit.a1600159 [Accessed 3 January 2019]. How scientists from a range of fields search for and read papers.

Primate Specialist Group Best Practice Guidelines. www.primate-sg.org/best_ practices/ [Accessed 3 January 2019]. A series of guidelines developed by the International Union for the Conservation of Nature's Primate Specialist Group to address critical issues in great ape conservation.

Many university libraries provide useful online guidance for how to search the literature.

11

Reading Journal Articles

Critical reading is an essential skill for scientists. As you design, conduct and report your study, you'll come back to the literature again and again, to find out more about particular topics. Reading takes time and can be daunting, but it gets easier with experience. Reading also teaches you what goes where in a paper. The more you read, the better you will write.

In this chapter I explain how to read articles, beginning with general advice, then providing questions to ask as you read each section of an article. Then I cover organising a reference collection and synthesising what you read.

11.1 GENERAL ADVICE ON READING

Expect to read key articles many times. You might read an article fast, initially, and then read it more carefully. You might also read an article for different reasons at different stages of your project. As you read other work, you may think differently about articles you read previously and need to re-read them.

Take notes as you read. Summarise each paragraph in one sentence. Journal articles often contain unfamiliar vocabulary. This specialised terminology is accurate and efficient, and aids communication between experts. As you read, look up words you don't understand and make your own glossary.

Primary research articles have a standard format, with some variation (Box 11.1). This means you can quickly find the information you want. Understanding one section of an article often requires reference to another part. Well-written articles guide you clearly through the logic of a study.

Box 11.1: Sections of a scientific report

Scientific reports have a standard format. Each section contains specific information. The headings vary, but are typically as follows.

Section	Content
Title	What the work is about
Abstract	The key points
Introduction	Background and why the authors did the work
Methods[a]	How the authors did the work
Results	The data
Discussion	What the authors think their findings mean
Literature cited	Sources cited in the text

[a] Some journals put methods at the end or in electronic appendices.

Authors vary in how clearly they write. When you find a well-written article, note why you found it easy to read and learn from this for your own writing. If you find an article very difficult to read, try another one on the topic. If you find them all difficult, then read a more general text, or a review paper, to understand the background to the topic before returning to the empirical literature.

11.2 THE TITLE, ABSTRACT, AND KEYWORDS

The title, abstract, and keywords of articles are freely available. There is an element of sales in the title and abstract of an article, so use them to decide whether to read further, not to assess the conclusions of a study. Never rely only on the abstract to evaluate a study.

11.3 READING THE INTRODUCTION

The introduction tells you why the authors did the work. It begins with the general context (the *big question*), then narrows to focus on the specific case study, describing the particular species studied, specific questions, aim, and how the authors addressed the aim. It ends with hypotheses and predictions, and sometimes with a preview of the significance of the results.

Read the introduction carefully, and summarise it for yourself, including:

- The big question
  - What is the problem this entire field is trying to solve?
  - Why is this problem interesting?
- The background to the study
  - What work has been done before to answer the big question?
  - What are the limitations of that work?
  - What do we not yet know?
  - What needs to be done to address this gap?
- The specific research questions the authors set out to answer
  - What are these questions?
- What the authors did to answer these specific questions
  - What hypotheses and predictions do they test? (these may not be phrased explicitly)

Reflect on the introduction. Is the article relevant to your question? If it is, then reflect on the rationale for the study. Do the authors review the literature comprehensively and fairly, or are they selective? Do you agree with the authors' assessment of the problem? Will their study fill the gap they identify? Are the hypotheses appropriate? Do the predictions arise from the hypothesis? Are the underlying assumptions correct?

11.4 READING THE METHODS

The methods should tell you exactly how the authors did the work with enough detail for another researcher familiar with the topic to repeat the study. They include technical details of how the study was conducted, the subjects, conditions, study design, data collected, and data analysis. You must understand the methods to assess the results and interpretation of the findings.

Read the methods carefully, and summarise them for yourself, including:

- The number of study subjects and how they were selected
- The size of the dataset and how the authors determined it
- The study design
- Exactly what the authors measured (the variables), and how

Reflect on the methods. Is the sample size adequate to address the question? Are the measures valid? Do the analyses test the predictions that the authors set out to test? We'll look at these questions in more detail in later chapters.

11.5 READING THE RESULTS

The results tell you what the authors found and present the evidence in the form of tables and figures.

Read the results carefully, paying careful attention to the figures and tables. Summarise the results for yourself, including:

- Do the results answer the specific questions posed in the introduction?
- Do the results support the predictions?
- How big are the differences between groups and how strong are the relationships reported? Don't focus only on the results of null hypothesis significance tests and p values – remember that the significance of a test is less important than the magnitude of the measures (Chapter 5).
- What do you think the results mean?

Reflect on the results. It's tempting to move on and find out what the authors think about their results, but don't. Instead, think about this for yourself before reading the authors' interpretation of their findings.

11.6 READING THE DISCUSSION

The discussion tells you what the authors think the results mean. It should include an assessment of any limitations. It covers whether the results support the predictions and puts them into the context of previous findings. It proposes possible explanations for the patterns observed and proposes future work to improve our understanding of the research question. The discussion usually ends with conclusions.

Read the discussion carefully and summarise it for yourself, including:

- What the authors think the results mean
- Any limitations of the study that the authors identify
- The main conclusions of the study
- The evidence the authors use to support those conclusions

- What the authors propose for future research
- Why are the conclusions important?

Reflect on the discussion. Do you agree with the authors' interpretation? Do the results show what the authors say they show? Don't assume that they do. Are there alternative ways to interpret the results? Do the authors compare their results to other studies, including contradictory findings? Do they offer reasonable explanations for any differences? Do you see any weaknesses that the authors missed? Don't assume the authors are infallible.

Don't accept the authors' conclusions without assessing them. Look at the evidence they present to support their conclusions. Have they overlooked, or downplayed, findings that don't support their hypotheses? Look at the reliability of the measures and the size and direction of the estimated effect, not just the significance. Do they overstate their findings?

It's okay to revise your opinion of the results when you read the authors' discussion. As a beginner, you probably will, but over time you will develop a more critical mind.

11.7 ORGANISING YOUR REFERENCE COLLECTION

Keep track of your reading and your notes. Archive digital copies of the articles you read and use reference management software to organise your reference collection. (There are very good free options.) Back this up. Start using reference management software from the beginning of your study. Don't wait until you have multiple folders of articles on your computer. Organisation will be important when you start writing up your work. Some reference management software also allows you to annotate articles electronically.

Take notes in your own words. If you copy directly from a source into your notes, then record this carefully. If you use copied text verbatim in later writing, this constitutes plagiarism (Section 3.1).

Look backwards and forwards from the article you've read. What literature does it cite? Who has cited it? Note new articles to find and read, and any new search terms that will be useful.

As you read, make a list of articles, each followed by a summary of the notes you made on it (an **annotated bibliography**). Describe the study briefly, identify the question, evidence, and conclusion, note the relevance to your own question, and assess the findings critically.

11.8 SYNTHESISING WHAT YOU'VE READ

Once you have read several articles on a topic, a question, a study species or system, or a method, write a summary of what you now know about it (Box 11.2). Synthesise what you have learned, concentrating on theory and evidence. Determine common themes and ideas, and evaluate the evidence for and against them. Seek patterns and potential explanations for differences among studies. Concentrate on the nature and size of the findings reported, rather than the statistical significance. Include definitions of important terms.

Use your literature review to identify questions that remain unanswered, and studies that need replication (Section 1.1) or extension to test the validity and generality of findings.

Box 11.3 reviews common errors in literature reviews and how to resolve them.

Many primatologists use a narrative review of the literature to justify a further study. However, this can easily be selective or biased. **Systematic reviews** go beyond this, using a thorough literature search

---

Box 11.2: Questions to ask yourself when reviewing the literature

Questions to ask yourself when synthesising what you have read include:

- What do we know about the topic?
- What are the major questions?
- What existing hypotheses address those questions?
- What evidence is there supporting, and against those hypotheses?
- What is unclear or contradictory?
- What do we not yet know?
- Which hypotheses still need to be tested, or tested further?
- What other lines of evidence would help to test the hypotheses?
- What confounding variables should you be aware of?
- Do we need a new study of the question? If so, why?
- What contribution can your proposed study make?
- What research methods are used to address the question?
- Are the research methods used satisfactorily?
- Are better methods now available?

Box 11.3: Common errors in literature reviews and how to resolve them

Common errors in literature reviews include the following issues:

Summarising Studies in Historical Order
Researchers often relate the history of research on a topic in a narrative form, for example *In the 1970s, author x did something . . . and in the 1980s, author y did something else . . .*. This historical approach can include material that is irrelevant to our current understanding (e.g. hypotheses that have long been refuted) and can be dull to read. Instead, concentrate on the major questions, current hypotheses, and the evidence base for each of them. Also avoid referring to research as recent when it is no longer recent.

Beginning with the Number of Studies of a Topic, Rather than Its Importance
Researchers often justify the need for a new study by citing the number of studies on a topic. However, this is not a good measure of the importance of a topic. There may be few studies of a question because it is not interesting, the phenomenon is new, or it is difficult to study. There may also be lots of studies of a relatively uninteresting question.

Summarising Each Article in a Separate Paragraph
Some reviews summarise each article in a separate paragraph, without synthesising the findings, often including irrelevant information about a study. Paragraphs beginning *In one study . . .* and *In another study . . .* are a sign of this. Such lists are a good start for a review, but they are not a review. Synthesise what you have read to determine common themes and ideas and evaluate the evidence for and against them. Cite sources more than once if they address different themes. Open each paragraph with the topic you address in that paragraph.

Listing Articles without Explaining What They Show
Some reviews contain lists of previous studies of a topic without reviewing what they found. Always include the relevant findings. Did the studies support or refute the hypothesis, and what was the effect size?

Box 11.3: (cont.)

Including Irrelevant Details

Some reviews include irrelevant details of the studies they review. For example, they might include unnecessary details of the methods. Similarly, there is no need to report the aim of a study in one sentence then the findings in the next. Just report the relevant findings, succinctly.

Stating That We Don't Yet Have a Complete Understanding of a Question

We will never have a complete understanding of a question because each study reveals further questions. Moreover, no individual study can hope to fully resolve a question, due to the need for replication (Box 1.2).

Concentrating on Statistically Significant Effects without Reference to Effect Sizes or Statistical Power

Don't simply list articles that did, and didn't, detect a statistically significant effect (a very common pattern). Instead, look at the size of the effects that studies report. Better still, conduct a meta-analysis (Section 11.8).

Assuming That There Will Be Only One Answer to a Question across Species

Some reviews highlight inconsistencies between tests of a hypothesis across case studies and suggest this is a problem. However, we expect case studies to differ and should seek patterns and potential explanations for differences between studies.

Relating the Literature to Your Study

A literature review should review what we know about a topic. Don't relate each point to your own study, and don't include a narrative of how you went about searching the literature.

with defined keywords to find all relevant studies, then summarising the findings. **Meta-analysis** is an excellent way to identify important variables and plan your study and is underused in primatology. A meta-analysis is a statistical analysis which combines the results of multiple studies addressing the same question, or very similar questions. It allows us to estimate the effect size (Section 5.7) with greater confidence, and to examine the factors which moderate the effect. Meta-analysis can also detect publication bias (Section 3.4).

11.9 CHAPTER SUMMARY

In this chapter, we've seen that:

- Critical reading is an essential skill.
- Reading gets easier with experience.
- We should expect to read key articles many times.
- Articles have a standard format.
- We shouldn't be distracted by the authors' claims when reading. Instead we should examine their reasoning and the evidence they present carefully, including the reliability of the measures and the size and direction of estimated effects.
- Reference management software helps us to stay organised.
- We use literature reviews to synthesise what we have learned.
- Meta-analysis is very useful to assess the current state of knowledge.

Once you have read the available literature on a topic and summarised what we know, you can identify what we don't yet know. This will help you narrow your research focus to a gap in our understanding that you may be able to fill. In the next chapter we look at how we formulate hypotheses, derive predictions from them, and design a study to test those predictions.

11.10 FURTHER READING

Gurevitch J, Koricheva J, Nakagawa S, Stewart G. 2018. Meta-analysis and the science of research synthesis. *Nature* 175: 175–182. A review of the benefits of meta-analysis in synthesising research.

Pain E. 2016. How to (seriously) read a scientific paper. https://doi.org/10.1126/science.caredit.a1600047 [Accessed 3 January 2019]. *Science Careers* blog. A collection of advice from scientists at different career stages.

Sayer EJ. 2018. The anatomy of an excellent review paper. *Functional Ecology* 32: 2278–2281. https://doi.org/10.1111/1365-2435.13207x. A very useful editorial on how to synthesise the literature, rather than simply summarising it.

Formulating Hypotheses and Predictions and Designing a Study

A clearly defined research question allows us to formulate hypotheses that propose possible answers to that question. From these hypotheses, we can then derive specific, unambiguous predictions that we can test with empirical data. Hypotheses and predictions serve to narrow down the infinite possibilities for data collection and determine the data we need to collect. Moreover, our predictions dictate the statistical tests we can use to analyse our data and draw conclusions, making careful thinking at this stage crucial to the success of a project.

In this chapter, I cover formulating hypotheses and predictions and explain that we often use proxies to test predictions. I describe how we can predict differences between groups, differences within subjects under different conditions, or associations between variables, and then how we extend this to predictions involving multiple variables. I revisit the problem of confounding variables, and end with how practical constraints influence the predictions we can test.

12.1 FORMULATING HYPOTHESES

Once we have narrowed our focus to a single research question, we can list potential answers to that question. These potential answers become our hypotheses and form a starting point for investigation.

A hypothesis is an educated guess as to what is happening, based on what we know. We base hypotheses on observations reported in the literature or our own observations. We need to read the literature to know what explanations other researchers have already proposed and the evidence that supports, or does not support, those hypotheses.

A good hypothesis is clear, precise, unambiguous, and as simple as possible. We must be able to reject a hypothesis based on evidence. In other words, hypotheses are testable and falsifiable.

If we know nothing about a phenomenon, then our research plan might be exploratory, aimed at generating hypotheses to explain it. However, in most cases in primatology, we can formulate hypotheses based on the existing literature. This literature doesn't need to be on our own study species; we can apply theory developed for other species to a new species, based on what we know of similarities and differences between the species.

12.2 DERIVING PREDICTIONS

Once we have formulated our hypotheses, we can use them to derive predictions. A prediction states a logical and observable consequence of a hypothesis. We can state hypotheses and predictions as *If the hypothesis is correct, then we predict that we will observe . . . .*

A hypothesis can give rise to many different predictions. Moreover, a specific prediction can be consistent with more than one hypothesis. We can test competing hypotheses that give rise to different predictions to rule out possibilities and determine the best explanation for the phenomenon we observe. However, much of primatology is more complicated than this. Hypotheses are often not mutually exclusive and the effect of one on a system may amplify or reduce the effect of another, or they may cancel one another out.

Deriving predictions takes careful thought but is essential to good study design. If other researchers have tested the hypothesis, scrutinise their reports to improve your own predictions.

As we saw in Chapter 5, the statistical relationships we predict take two forms, depending on the types of variables involved. We might predict a difference between two or more groups, for example a difference between species, sexes, or experimental treatments. Alternatively, we might predict a relationship between two quantitative measures – an **association** or a **trend**.[1] For example, we might predict a positive relationship between body mass and body length among adult animals.

Box 12.1 describes common problems with hypotheses and predictions and how to resolve them.

---

[1] Some researchers also use the word *trend* to describe a pattern with a p value close to, but not smaller than, 0.05.

Box 12.1: Common problems with hypotheses and predictions, and how to resolve them

Researchers often confuse hypotheses and predictions, but the difference is important. Moreover, one or both are often missing from a report, but we should include both. Practice identifying hypotheses and predictions in the literature you read; you'll notice that not all articles get them right.

Stating a Hypothesis but No Predictions

Some researchers state their hypothesis, but don't follow it with the predictions they used to test it. Check that you state both clearly.

Stating the Prediction as a Hypothesis

Some researchers state the prediction – what they expect to observe – as their hypothesis. Ensure that your hypothesis is a potential explanation for a phenomenon, not the specifics of what you expect to observe if that explanation holds.

Stating Predictions but No Hypothesis

Some researchers state what they expect to observe, but don't state the hypothesis that gives rise to their prediction(s). If you have predictions, then there must be an underlying hypothesis. Adding *because* at the end of the prediction helps to identify the hypothesis. Then reverse the statement to obtain an *If ... then ...* statement of the hypothesis and prediction(s).

Confusing Methods with Hypotheses

Some researchers state their method, rather than a hypothesis, perhaps because they misunderstand the *If ... then ...* formulation of the hypothesis and prediction. Resolve this by checking that the *If* statement is an explanation, not what you intend to do.

Setting Non-Mutually Exclusive Hypotheses as Competing

Some researchers set up two hypotheses for a phenomenon as though they are mutually exclusive when they are not. This is particularly common when testing proximate and ultimate explanations for the same phenomenon (Section 1.1).

Incorrect Language

We *test* a hypothesis; we don't *verify* or *confirm* it. Our data may *support* a hypothesis or *reject* it, but we can't *prove* it.

12.3 MEASURING INVISIBLE THINGS

Predictions should be as specific as possible, and we must be able to collect data to measure them. In other words, predictions describe the variables we need to measure. However, we can rarely measure the variable we wish to study directly. For example, we can't measure concepts such as *cognition*, *risk*, *stress*, *quality*, or *health* directly. Nor can we measure many physiological variables without invasive tissue sampling. We may not be able to measure a variable within the timeframe of a study. For example, behavioural ecologists can rarely measure lifetime reproductive success, and conservation projects may not be able to measure the influence of an intervention on population trends in their target species.

Where we can't measure the outcome variable itself, we measure a variable we hope is related to the concept we would like to measure and use that **proxy** variable to infer patterns we cannot observe. Proxies involve assumptions and our measurement tools can be blunt, so we must consider and justify the assumptions we make carefully. It's not sufficient to copy what other researchers have done. We'll go into this in more detail in Chapter 14.

Examples of inadequate proxies are using time invested to measure the quality of work, the number of words written to measure progress when writing, or journal impact factors to measure the quality of academics (Box 10.2).

We must examine the validity of a proxy when interpreting our findings and drawing conclusions. Using proxies makes it difficult to reject a hypothesis if we find no support for a predicted difference or relationship, because it is easy to explain why the proxy is inadequate if we don't find the predicted relationship.

12.4 TESTING FOR DIFFERENCES BETWEEN CATEGORIES

If we predict a difference between two or more categories, we can test this by testing for differences between groups of subjects, or for differences within subjects.

12.4.1 Testing for Differences between Groups

If we test for differences between categories by measuring separate groups of subjects for each category, the data in different categories are independent of one another, the predictor variable varies *between* groups, and we want to know the size of the difference. This is a **between-group** or **independent measures** design. For example, we might want to compare measures for males and females. Between-group designs include comparisons of an experimental group with a **control** group that does not receive the treatment.

Simple between-group designs measure each subject once only. However, unwanted variability (**noise**) within a group can be large, making the estimate of the effect size imprecise and potentially obscuring any relationship. We can improve this by using a large sample size. We can also reduce variability within groups by comparing like with like and keeping confounding variables (Section 5.3) as equal as possible. For example, we can select subjects that are of the same sex and age class (Box 7.1) to reduce the influence of these variables on our measures. However, doing this limits our ability to generalise from our results to the other sex or other age classes.

If we assign subjects to groups at random, manipulate the predictor variable experimentally and detect an influence on the outcome variable as a result, we can conclude that change in the predictor variable causes change in the outcome variable. However, if we use pre-existing groups, rather than assigning subjects to groups randomly, we cannot draw conclusions about cause from a between-group design, because we cannot rule out the influence of pre-existing differences between the groups that cause the difference (Section 5.3). We may have no choice about this. For example, a comparison of males vs. females must be a between-group design because males and females are pre-existing groups.

We draw a predicted difference between two or more categories by plotting predicted summary data (the count, median, or mean values, Box 5.1) for each category on the y-axis, with category on the x-axis.

12.4.2 Testing for Differences within Subjects

If we test predictions by measuring each subject more than once, under different conditions, a single group of subjects contributes data for all conditions (or treatments) and the comparison we're interested in is *within* the individual. For example, we might measure a set of animals in different seasons, in different experimental conditions, or as they develop. This is a **within-subject** or **repeated measures** design. Where there are two conditions, this is a paired design.

Within-subject designs account for variation in individual response and look only at the size of the difference in scores between conditions for the *same* individual. This design accounts for the fact that some subjects may score high in all conditions while others score low in all conditions by matching the scores for each subject with the other scores from the same subject. In other words, it accounts for the fact that the data for each subject are related (or **dependent**[2]) because they are for the same individual. Our sample size is the number of cases, not the number of data points. This has serious implications for statistical analysis, which must account for the dependence between the measures of an individual case (more on this in Chapter 15).

Within-subject designs are more sensitive than between-group designs and yield a more precise estimate of the difference between conditions than independent groups, particularly if values are highly correlated between conditions. Thus, we can use a smaller sample size than in a between-group design. (We'll look at sample size in more detail in Chapter 16.)

If we're not careful, the first condition in a within-subject design may affect the second (**carry-over effects** or **order effects**). For example, animals may initially show great interest in a novel stimulus, but then quickly habituate to it and ignore it. We can avoid such effects by **counterbalancing** the order of testing. For example, we can randomly assign separate groups of subjects to conditions in a different order (e.g. comparing two treatments, A and B, with one group of subjects receiving Treatment A followed by Treatment B and the other receiving Treatment B followed by Treatment A).

To plot a predicted difference in a within-subject design, we use the same plot as for testing for differences between groups and add lines to link matched data points.

12.5 TESTING FOR ASSOCIATIONS BETWEEN QUANTITATIVE VARIABLES

Simple associations between two quantitative (ordinal or continuous) variables (Section 5.2) can be positive, where one variable increases as the other increases, or negative, where one variable decreases as the other increases. Relationships can be a straight line (**linear**), with change in one variable directly proportional to that in the other, or **non-linear**, like the relationship between body mass and age.

---

[2] Don't confuse dependent and independent *data* with dependent and independent *variables* (Section 5.3).

As with predicted differences between categories, data in predicted relationships can be independent (one measure for each variable per subject) or dependent (two or more measures per subject). Again, we need to incorporate any dependence in our data in our statistical analysis.

We draw a predicted relationship between two quantitative variables by plotting one variable on the x-axis and the other on the y-axis of a bi-plot. Where there is no clear causal explanation between the two variables, we plot either of them on the x-axis. Where we predict that change in one variable is responsible for change in the other, we plot the cause on the x-axis and the predicted outcome on the y-axis. This does not mean that we can infer causation if our data support the prediction, only that we hypothesise it.

12.6 PREDICTIONS INVOLVING MORE THAN ONE PREDICTOR VARIABLE

The effects of two or more predictors on an outcome variable can be **additive**. In other words, the effects of one predictor on the outcome do not affect those of the other. For example, we might predict that both sex and body length influence body mass.

We might also predict that the relationship between a predictor variable and the outcome variable depends on the value of another variable, in an **interaction**. For example, the relationship between dominance rank and testosterone in male primates may depend on whether the dominance hierarchy is stable, or the reaction to a stimulus may depend on whether the subject has already been exposed to it.

We can plot an interaction with the outcome variable on the y-axis (e.g. body mass), the first predictor on the x-axis (e.g. age), and different lines to represent values of the second predictor (e.g. females and males). If both predictor variables are continuous, we can plot the outcome variable on the y-axis, the first predictor variable on the x-axis, and the second predictor variable on the z-axis in a three-dimensional surface plot, or divide one of the continuous variables into levels (e.g. low, medium, and high) and plot the lines for each level on a two-dimensional plot of the two other variables.

We can combine between-group and within-subject design in a **mixed design**. This is very common in primatology, as it allows us to make the most of complex datasets. For example, you might aim to collect multiple measures of an outcome variable for a set of focal subjects, nested with groups, over several years, often with gaps in the dataset where you can't obtain the data. These situations require careful statistical modelling and you are likely to need advice on this to avoid collecting data that you cannot use to test your predictions (we'll learn more about planning data analysis in Chapter 15).

12.7 CONFOUNDING VARIABLES

We need to consider potential confounding variables that influence the outcome variable and may prevent us from observing the effect of our predictor variable (Section 5.3). For example, if one observer is responsible for data collection on one group, and another observer is responsible for data collection on a second group, any difference between the two groups may be due to differences in the observer, rather than to real group differences (**observer bias**). Similarly, **batch effects** can influence the outcome of experiments. For example, if we run all the tests for one group of subjects on one day and all the tests for another group on another day we cannot rule out the effect of differences in conditions between days on the outcome of our comparison. The solution to this is to randomly assign subjects to groups.

12.8 PRACTICAL CONSTRAINTS

As we saw in Chapter 9, our chosen research question should enrich our understanding of primates and we should be able to address it in the time we have available and with the resources at our command. Switching to a different study taxon or a different site can help to resolve constraints and allow us to address a question. In other cases, we need to reconsider the questions we address.

Ideally, we select the best species and conditions to test our hypotheses and predictions. However, in reality, we are limited by what is available, and by practicalities. We may be limited to a particular study location, and the conditions there may limit our options in terms of study design. Funding is also often a major constraint.

For studies that rely on museum specimens, there may be few (or no) specimens available in collections, we may not have access to them, or we may not be able to use the destructive sampling needed to obtain samples. For studies involving live animals, there may be no habituated animals available or a study site may be fully booked. Similarly, there may be few (or no) captive animals available, or we may not be able to gain access to them. We may be unwilling (or unable) to use invasive sampling, limiting our ability to address physiological questions. If we hope to use data from the literature, there may be too few data available to test a prediction, or there may be insufficient evolutionary transitions for a comparative analysis.

The availability of study specimens, animals, or data and the feasibility of data collection place practical constraints on the predictions we can test, and thus the research questions we can address. These constraints make the process of formulating a research question, framing hypotheses, and deriving predictions an iterative one, until we finally arrive at a question we can hope to answer.

12.9 CHAPTER SUMMARY

In this chapter, we've seen that:

- A clearly defined research question allows us to formulate hypotheses that propose possible answers to it.
- We can formulate hypotheses based on the existing literature for most research questions in primatology.
- We derive predictions from hypotheses to test their validity with empirical data.
- We use proxies to measure invisible concepts.
- Predictions are either differences between categories or associations between variables.
- Well-formulated predictions tell us exactly what data we need to collect to test them and how to analyse the outcome.
- We should be able to draw our predictions in a simple plot.
- Data can be independent (one measure for each variable per subject) or dependent (two or more measures per subject).
- Predictions involving multiple variables can become complicated and need careful thought.
- The process of formulating a research question, framing hypotheses, and deriving predictions is iterative; we keep revising our ideas until we arrive at a set of testable predictions.
- Feedback is important to make sure our ideas make sense and are feasible.

The next chapter looks at different ways of testing our predictions, using observations or manipulations.

12.10 FURTHER READING

Barnard C, Gilbert F, McGregor P. 2011. *Asking Questions in Biology: A Guide to Hypothesis Testing, Experimental Design and Presentation in Practical Work and Research Projects*. 4th edn. Harlow: Benjamin Cummings. An excellent introduction to the research process in biology. Chapter 2 covers the art of framing hypotheses and predictions.

Field A, Hole G. 2003. *How to Design and Report Experiments*. London: Sage Publications Ltd. Chapter 3 includes simple and more sophisticated experimental designs.

Hailman JP, Strier KB. 2006. *Planning, Proposing, and Presenting Science Effectively*. Cambridge: Cambridge University Press. Chapter 1 includes formulating hypotheses and devising testable predictions.

Stamp Dawkins M. 2007. *Observing Animal Behaviour: Design and Analysis of Quantitative Data*. Oxford: Oxford University Press. Chapter 2 covers formulating hypotheses, with a specific focus on observational studies of animal behaviour.

13

Observing and Manipulating

To test our predictions, we either observe natural variation and use statistical analysis to account for variation in other variables (**covariates**) to estimate the influence of the variable we're interested in, or we control variation in a planned manner, by manipulating one predictor variable and holding others constant.

In this chapter, I introduce good study design, and then explain that although it is easiest to describe observations and experiments separately, they lie at opposite ends of a continuum of researcher-imposed control on a study system. I introduce the concept of validity and review the strengths and weaknesses of observation and manipulation. I finish with a reminder to consider whether a study is feasible.

13.1 GOOD STUDY DESIGN

Good study design is the essence of good science. It is impossible to observe and record everything systematically, so determining what to measure is key.

Many problems with study design arise when researchers focus on logistics or feasibility rather than on how to address their aims, apply methods they have been trained in or read about without understanding why, or venture into new disciplines without understanding the underlying theory. Of course, feasibility is important, but it is crucial that our study design addresses our aims. Where this proves impossible, we must revise our research question to one that we can address within the constraints we face. Changing our study taxon or location can also help.

13.2 A CONTINUUM OF CONTROL

Observations and manipulation are both scientific procedures undertaken to make a discovery (**experiments**), but we typically use the word experiment for manipulative studies. Experiments are sometimes viewed as more rigorous than observations, but all methods are limited and we need multiple lines of evidence to be confident in an explanation for a phenomenon (Section 1.1).

Observations and controlled experiments are the two ends of a continuum of researcher-imposed control on a study system. Between the two lie **natural experiments**, with no manipulation, but where variables vary systematically; **quasi-experiments** with systematic sampling of unmanipulated predictor variables; and **weak experiments** where we manipulate one or more variables, but don't hold other parameters constant. Our ability to infer causation increases as control increases.

13.3 VALIDITY

The results of a study may be true, but they may also be erroneous, resulting from the way the study was designed or conducted (an **artefact**). A test is valid if it measures what it claims to measure.

Internal validity is the degree to which we can conclude that our findings are due to the predictor variable and not some other variable. In other words, it is the degree to which we can conclude that there is a causal relationship between the outcome and the predictor variables.

Spurious relationships can arise if changes in the outcome variable are due to a confounding variable, rather than to our predictor (Section 5.3). We can improve internal validity with careful study design, including careful selection of samples, controlling for all extraneous variables, using standardised methods, counterbalancing the order of testing such that different subjects receive treatments in a different order (Section 12.4.2), and using blind protocols where possible (Section 3.3).

External validity is the degree to which we can generalise the results of a study to other subjects and other settings. We can improve external validity by selecting a representative sample, setting experiments in a more natural setting, and choosing procedures carefully.

Internal and external validity often trade-off against one another. Efforts to increase internal validity may constrain the generalisability of our results, while efforts to improve external validity may limit our ability to account for extraneous variables.

Ecological validity is the degree to which the study approximates the real world. Artificially controlled and constricted environments reduce ecological validity. Although they are related, ecological validity is distinct from external validity. A study with high ecological validity can have low generalisability, and vice versa.

13.4 OBSERVATION WITHOUT MANIPULATION

In an observational study, we measure natural variation in a system under different conditions and test for connections between variables using statistical models. Some studies can only be observational, because they rely on specimens that exist in museum collections, or existing data in the literature. For example, we might test whether aspects of tooth morphology are related to diet, using museum specimens, or we might collate data from the literature to compare primate testes size with the number of males that mate with a female during her fertile period across species. Other observational studies involve behavioural observations or measurements of primate morphology and physiology in a variety of captive and wild situations, without experimental manipulation. For example, we might test the influence of maternal traits on offspring growth patterns and physiology, seasonal variation in diet on the primate gut microbiome, or nocturnal luminance on activity patterns.

We can also make carefully matched control observations to compare with the behaviour we observe following particular events. For example, we might compare an individual's behaviour after an episode of spontaneous aggression with their behaviour at another time, matched as closely as possible to the first observation, but with no preceding aggression.

Observational research is generally more feasible than experimental manipulation. It measures the natural range of variation in our animals (e.g. in their behaviour or physiology) and their environment, and thus has high ecological validity. With a carefully selected sample, we can generalise from the results to other situations and subjects, so our study also has good external validity. Moreover, if things go well, we can obtain a lot of data.

Observational studies need careful thought and planning to effectively isolate the effects of correlated variables. We can't assign subjects to groups randomly, so groups almost certainly differ in more than just the variable we're interested in. These confounding variables (Section 5.3) limit the internal validity of observational studies. A carefully designed sampling strategy can account for confounding variables statistically, at least to some extent. This requires a good sample size, and the structure of our data can limit the comparisons we can make. For example, if two predictor variables are highly, or perfectly, correlated with one another, either positively or negatively (i.e. they **covary**), then we cannot draw conclusions about either of them in isolation (Box 13.1 gives examples).

As we saw in Section 5.3, if we observe a relationship between two variables (a correlation), we cannot necessarily conclude that they are meaningfully related to one another, or that change in one variable causes change in the other (causation). Careful design of observations can help to disentangle cause and correlation. For example, establishing a temporal relationship between variables can support a hypothesis of causation. For A to cause B, A must precede B. However, this is also not sufficient to conclude causation. Although we can hypothesise about the direction of cause and effect based on observations, we cannot establish that the predictor variable is responsible for any observed changes in the outcome variable, because we cannot rule out other explanations (alternative hypotheses).

A further disadvantage of observational studies is that we may rarely or never see what we want to measure.

13.5 MANIPULATIVE RESEARCH

In a manipulative study, we deliberately vary the predictor variable in a controlled manner, holding all other variables constant, to determine how it affects the outcome variable. We either apply a treatment to one group and compare it with an untreated control (between-group design, Section 12.4.1), or we apply each treatment to the same group of subjects (within-subject design, Section 12.4.2). For example, we might broadcast a sound (e.g. a conspecific vocalisation) from a hidden speaker, present olfactory or visual stimuli, or present an object (e.g. a model predator) and measure a study animal's response. We can also test animals using tasks designed to assess their cognitive skills.

The design and conduct of manipulative studies require a deep understanding of the study animals to understand the questions we can realistically address and enable us to design appropriate experiments.

Unlike in observational studies, we can assign subjects randomly to some conditions in manipulative research, although we can't assign them to age-sex classes, or other intrinsic characteristics. Control over variation and confounding variables allows us to test the causal relationship between the predictor and outcome variable, and whether manipulation of the predictor variable caused the effect on the outcome

Box 13.1: Problems with confounding variables and how to resolve them

Many problems with study design involve confounding variables (Section 5.3). If two predictor variables are highly or perfectly correlated, then we cannot draw conclusions about either of them in isolation. For example:

- If we compare the traits of one group in one habitat with those of a second group in a second habitat, we cannot draw conclusions about the influence of habitat on the traits we measured, because any differences may be due to differences between groups. To resolve this, we would need to measure multiple groups in each habitat, to isolate the influence of group from that of habitat.

- Similarly, if we compare the traits of one species in one habitat with those of a second species in a second habitat, we cannot draw conclusions about the influence of habitat on the traits we measure because any differences may be due to differences between species. To resolve this, we would need to measure multiple species in each habitat, to isolate the influence of species from that of habitat.

- If we compare one age class in one species with another age class in another, then any effect of species is confounded with the effect of age class. To resolve this, we would need to study the same age class(es) in each species.

- If we use different methods to measure different groups, any differences we detect may be due to the difference in methods, not to differences between groups. To resolve this, we must use the same methods for all study groups.

- If we measure different groups at different times (e.g. in different seasons or years), we cannot conclude that differences between groups are not due to variation across time (e.g. seasonal or inter-annual variation in ecology). To resolve this, we should measure groups at the same time.

- If we measure the same group before and after a treatment, any differences we observe may be due to the passage of time. To resolve this, we can include control groups that do not receive the treatment, to isolate the effect of the treatment from that of time.

Box 13.1: (cont.)

These are common errors in primatology and authors often attempt
to justify them on logistical or practical grounds, or by citing
studies that make the same error. However, such justifications
don't resolve the problems of inference.

The ideal solution is always to improve our study design to isolate
the predictor variable from confounding variables. In some cases,
we can achieve a partial solution by testing the influence of
confounding variables separately. If this influence is smaller than
that of the variable we're interested in, then we can tentatively
conclude that our variable may cause an effect.

FURTHER READING

Leavens DA, Bard KA, Hopkins WD. 2017. The mismeasure of ape social
cognition. *Animal Cognition*. A thorough treatment of confounded two-
group comparisons in comparative psychology. https://doi.org/10.1007/
s10071–017-1119-1.

variable. In other words, experiments achieve high internal validity by
controlling for confounding variables. To accomplish this, we need test
groups that are alike in all variables other than the manipulated variable. This requires careful thought and we should be alert to the possibility of confounding variables (Section 5.3).

It is easier to document non-events in manipulations than in
observations, for example, when animals do not detect a model predator. We can also test subjects' responses beyond the range of natural
variation, for example by studying animals in novel situations, or by
simulating encounters that are difficult to observe. However, such
manipulation may not be ecologically valid.

The disadvantages of manipulations are that they are usually less
feasible than observations and can be unethical, in which case we should
not pursue them (Chapter 2). Moreover, experiments may be so carefully
controlled that we cannot generalise from them. For example, the sample
may not be representative (low external validity) and the results may not
generalise to other settings (low ecological validity). In other words,
results showing that manipulation of a predictor variable can influence
an outcome variable do not mean that it does so under natural conditions. We can improve this by using the natural range of variation in a

predictor variable and conducting experiments in as natural a setting as possible (field experiments), but this is likely to reduce control over other parameters and can be very time-consuming.

13.6 PRACTICAL CONSTRAINTS

This is another time to stop and consider whether the study design we have in mind is feasible. It's important to seek feedback from fellow researchers (Box 13.2) and people at potential field sites, research stations, laboratories, collections, and zoos (Box 13.3). For example, animal keepers and field teams know the animals they work with extremely well and will be able to advise on what sort of observations and experiments are possible. Pilot studies also help (more on this in Chapter 17).

13.7 CHAPTER SUMMARY

In this chapter, we've seen that:

- Good study design is the essence of good science.
- Poor study design can lead to a great deal of wasted time and effort.
- We can control variation in one or more predictor variables to reveal the effect of each one on an outcome variable statistically or experimentally. Each has advantages and disadvantages.
- Observation and manipulation are the two ends of a spectrum of experimental control.
- Internal validity (our ability to draw conclusions about causal relationships) often trades off against external validity (our ability to generalise from our results).
- Observations have high ecological validity and can have good external validity. They often have limited internal validity.
- Manipulations have high internal validity but may lack external validity and have low ecological validity.
- Observation and manipulation complement one another, and we use each to refine the other.
- We should consider how practical constraints limit what is possible when designing a study.

In the next chapter, we look at choosing appropriate methods to measure the variables we need to test our predictions.

Box 13.2: Identifying and contacting experts

There are many points in a study where you need advice from experts, including the practicality of testing your hypotheses and predictions, appropriate methods, and peer review of proposals and draft reports. You may be lucky and have good mentors in your home institution. However, not everyone has easy access to expert advice, and we all need to seek expert advice from people we don't know at some stage.

How Do I Find an Expert?

You can identify experts by reading the literature and noting the authors of recent high-quality work on the topic you're interested in (Chapter 10). Not everyone working on a topic is a true expert, so check the work carefully.

Social Media

Twitter is a good way to make initial contact with other primatologists and can make an introduction at a conference or over email easier. You can also ask questions on social media, but make sure to do your research to avoid asking a question that you can easily answer for yourself.

Email Etiquette

Address the person you're writing to as Dr [family name] or Prof [family name], as appropriate. Don't assume the person is male. Customise your request to the person you're writing to. If he or she has a website, read it carefully. Explain your shared interests and how your work relates to the person's interests. Don't explain how the person's work is relevant to yours. You're asking a favour, so prioritise that person's work. Be polite and respectful.

Persevere if you don't get a response. A minority of researchers will be rude, arrogant, or unconstructive, but most are helpful. If you're unlucky, move on to another possible source of help. If you receive no response, send a polite follow-up email after about a week.

If you establish a mentoring relationship, read your mentor's work.

Box 13.3: Contacting potential study sites, collections, and laboratories

It's important to contact people at potential sources of data early in the design of a study and to consult them about your plans. This includes people responsible for field sites, zoos, animal sanctuaries, collections, laboratories, and databases.

When you contact a potential research site, include your name, position, institution, and contact information; the name of your mentor, if appropriate; and brief details of your request, with the data you wish to collect, the dates, and a short proposal. Offer to meet in person, if possible, or to video-chat to discuss your projects. Some places may have research request forms.

Facilities and the people who work at them are not usually there simply to facilitate our research, so we need to fit in with them, not the other way around. Respect the priorities of your prospective research site. You are asking them to help you and you may be adding to their usual workload. We must be willing to work within the constraints they set out and be flexible. For example, many zoos welcome researchers but their priorities are the visitor experience, animal welfare, and management. Animal sanctuaries may be interested only in research that benefits their animals. Collections may close for a time.

When we apply to work somewhere or with someone, we are often proposing a collaboration (Section 2.9). Ask about terms of use, including site or bench fees and co-authorship. You may need to contribute to long-term core data collection at a field site, provide a copy of any data you collect, and leave a duplicate of any samples. Also ask for a site's code of conduct (Section 4.11).

Ask about the site's research priorities and ongoing projects. Space can be limited at a field station, facility, or collection, and the potential research site may have competing demands. For example, other researchers may be studying the same animals or material or using the same datasets.

Explain your aims and proposed methods and ask about feasibility, including which methods will work, and which will not. Ask about any facilities you need (e.g. freezer space, power, research assistance) and discuss logistics, which may include permits, transport, and accommodation.

Box 13.3: (cont.)

People can be very protective of their research site and material, reflecting the hard work they have invested to establish a site, collection, or database. People may also have negative preconceptions of researchers as arrogant. Be respectful, understand other people's points view, don't assume that you know best, and don't expect others to accommodate you. A single badly behaved researcher can cause havoc at a research site. Specimens may be fragile. Destructive sampling requests, in particular, require detailed justification. Having said all that, don't be afraid to ask what's possible.

Finally, when you agree the remit of a study, ensure that the agreement reaches all the decision makers, to avoid later complications.

13.8 FURTHER READING

Altmann J. 1974. Observational study of behavior: Sampling methods. *Behaviour* 49: 227–267. https://doi.org/10.1163/156853974X00534. Section 1B covers manipulative and non-manipulative research.

Field A, Hole G. 2003. *How to Design and Report Experiments*. London: Sage Publications Ltd. Chapter 3 gives more detail on experimental designs.

Janson C. 2012. Reconciling rigor and range: Observations, experiments, and quasi-experiments in field primatology. *International Journal of Primatology* 33: 520–541. https://doi.org/10.1007/s10764–011-9550-7. Covers how and why we seek to control variation in predictor variables, experiments and observations as endpoints of a continuum of control, and the benefit of generalised linear mixed models in primatology.

Sagain R, Pauchard A. 2012. *Observation and Ecology: Broadening the Scope of Science to Understand a Complex World*. Washington, DC: Island Press. Documents the shift in approach from observation and discovery to a heavy focus on experimental manipulation in ecology and makes the case that observational research is critical.

Stamp Dawkins M. 2007. *Observing Animal Behaviour: Design and Analysis of Quantitative Data*. Oxford: Oxford University Press. Chapter 5 covers observational study design.

Zuberbühler K, Wittig RM. 2011. Field experiments with nonhuman primates: A tutorial. In: Setchell JM, Curtis DJ (eds.). *Field and Laboratory Methods in Primatology: A Practical Guide*. Cambridge: Cambridge University Press. Reviews experimental designs and stimuli used in field experiments with wild primates, including common problems.

14

Choosing Measures

Good research design includes choosing what to measure and how to measure it. We can't measure everything. Fortunately, clear predictions dictate the measurements we need to make to test them. The next question is how to obtain those measurements.

This chapter provides general advice on methods, then covers the importance of the validity, accuracy, and sensitivity of the measures we use. I end with a reminder that methods must also be feasible.

14.1 GENERAL ADVICE ON METHODS

Methods are dictated by the variables we need to measure to test our predictions. Anything else we measure is an optional extra. If our methods don't measure what we need to measure (i.e. they are not a good proxy), then the data we obtain won't answer the research question. We can fix this at the planning stage. Sadly, however, researchers often collect a lot of data before they realise that their methods don't adequately measure the variables they wish to measure.

List the variables that you need to measure. Assess the potential methods available to measure them and summarise the state-of-the art for doing so. Use established methods wherever possible to facilitate comparison with published studies, but only if they are appropriate. Critically evaluate the methods used in published studies. What limitations do the authors acknowledge? Can you see any other limitations? Can you improve on their methods? The most appropriate way of measuring a variable of interest may not be the most high-technology method available.

Look for review articles that describe best practice and highlight common problems. Some methods are debated, and you may find no consensus in the literature, even among the experts. Beware of echo-chambers in which authors and reviewers from a limited pool fail to spot problems with methods.

Explore your options and contact experts to ensure that you use appropriate methods (Box 13.2). Get trained by experts if possible. You want to lead the way, not repeat other researchers' mistakes. Don't assume that published, or even standard, methods are appropriate or valid and don't simply copy what others have done, something you have learned, or a method someone suggests, without first learning the scientific basis of the method.

You can convert data from a higher resolution to a lower resolution by splitting continuous data into categories, but you can't do the opposite, so collect data at the highest practical level of resolution relevant to your study. Continuous variables allow more sophisticated statistical analysis than discrete variables, but the level of detail will depend on the prediction you wish to test.

If your methods are from other disciplines, it is crucial to collaborate with experts in those disciplines. Interdisciplinary collaboration can be difficult initially and takes us out of our comfort zone but is extremely rewarding (Section 2.9). It brings an opportunity to discover new approaches to existing questions and identify new questions to address.

Be critical and consider possible problems, but don't get completely hung up on achieving perfection; it's not possible.

14.2 ARE MY MEASURES VALID?[1]

The key question for any measure is whether it tells you what you want to know. In other words, does it measure what you want to measure? As we saw in Section 12.3, we often use proxies to measure the unmeasurable. A proxy variable must be strongly correlated with the target variable. This is key, but often overlooked.

14.3 ARE MY MEASURES RELIABLE?

Reliability is the repeatability or the consistency of our measures. In other words, it tells us whether we will obtain the same or a very similar

---

[1] Don't confuse the validity of your measures with the validity of your study design (Section 13.3).

value if we measure the same thing again under the same conditions. It doesn't tell us whether the results are correct. Unreliable measures introduce unwanted variability due to **random error** into a dataset, making it harder to detect a signal of any effect. We can reduce the effect of random error by increasing the sample size.

14.4 ARE MY MEASURES ACCURATE?

Accuracy is the certainty in a measurement with respect to the actual value. The difference between what we can measure and the actual value is the **measurement error**. Inaccuracy leads to **systematic bias**, which may be okay if you don't need to know the exact value of your measure.

Measures can be repeatable, but not accurate. They can also be accurate but not repeatable. They can also be neither, or both.

Consider sources of bias and how you can measure them.

14.5 ARE MY MEASURES SENSITIVE?

Sensitivity is the smallest amount of change that your measure can detect. **Resolution** is the smallest change in the underlying quantity that produces a response in the measurement. The resolution of a measure must be sufficient to detect the amount of change we are interested in.

We also need measures that provide a broad spread of scores. Some measurements suffer from floor or ceiling effects, which prevent accurate data interpretation. A **floor effect** occurs when there is a lower limit (a floor) to the data values you can measure reliably and results in bunching of scores at the lower limit. A **ceiling effect** occurs where there is an upper limit (ceiling), and scores bunch at that upper limit. Both reduce variability.

14.6 ARE MY MEASURES FEASIBLE?

The feasibility of methods includes logistics and cost. Can you collect the data you need? In other words, can you obtain the samples you need, in sufficient numbers? This might be behavioural samples, faecal samples, urine samples, samples of primate foods, measures of habitats, measures of specimens, or data from the literature. If you need to store samples, can you do so appropriately? Can you transport them? How and where can you analyse them? Can you conduct the experiments you wish to conduct, with sufficient replicates? Do you have, or can you get, the necessary equipment, consumables, and reagents? If not, are there cheaper methods that will work?

Discussion with people at potential study sites and with experts in your field (Box 13.2) will help with these questions. A pilot study is invaluable (more on this in Chapter 17).

14.7 CHAPTER SUMMARY

In this chapter, we've seen that:

- Our predictions dictate the measurements we need to make.
- Researchers often invest a lot of time and energy in data collection before they realise that their methods don't measure the variables they wish to measure.
- We should evaluate the methods of published articles critically.
- We should use established methods to facilitate comparisons but only if they are appropriate.
- We shouldn't choose a method simply because it's familiar, or because someone else suggested it, without evaluating it carefully to be sure it measures what we wish to measure.
- We usually need training in methods.
- Collaboration is crucial when we venture into other disciplines to avoid naïve errors.
- Measures must be valid, reliable, accurate, sensitive, and feasible.

Box 14.1 addresses common problems with methods and how to resolve them. In the next chapter, we look at how we use statistical analyses to test our predictions.

Box 14.1: Common problems with methods and how to resolve them

All of the problems I list here can be resolved by ensuring that your methods are appropriate. You can achieve this by reading the literature to understand the methods you need to measure the variables you want to measure, validating methods carefully, and collaborating with experts when venturing into a new discipline.

Box 14.1: (cont.)

Problem	Solution
Using non-invasive endocrine assays without validating them analytically and biologically (e.g. using a kit without validating it). Failing to test whether and how storage affects hormone concentrations.	Use fully validated methods for non-invasive endocrinology. Collaborate with an expert in non-invasive endocrinology.
Not treating samples in exactly the same way and introducing a batch effect that confounds the data.	Treat all samples in exactly the same way. If this is impossible, test whether differences in treatment influence your results and correct for this if they do.
Using a storage medium or method that precludes the analyses you want to do.	Check your laboratory methods when planning sampling.
Studying parasites without expert understanding of parasitology. For example, non-experts might assume that microorganisms detected in faecal samples are pathogenic (they may also be beneficial, making simple counts of species richness dubious); misuse standard terms in parasitology; employ sampling regimes that may not measure what we hope to measure (e.g. the number of parasite eggs shed in faeces may not be a good indicator of infection because shedding varies over time); or wrongly identify gastrointestinal parasites to species level based on egg morphology. (This is particularly relevant to studies comparing parasites across hosts.)	Collaborate with expert parasitologists. Define exactly what you wish to know and validate methods to be sure that they measure what you hope to measure. Redefine your question if you cannot measure what you wish to measure. Use molecular methods and larval culture to identify parasites to species level.

Box 14.1: (cont.)

Adopting phenology methods already used at a site, at other sites, or in the literature without considering what you need to measure.	Begin with what you want to measure and design your methods accordingly. Collaborate with a botanist.
Assessing dyadic relatedness using a limited number of genetic markers and without a deep pedigree.	Understand the relationship between genetic and pedigree relatedness. Collaborate with a geneticist.
Using questionnaires, rating scales, and closed questions (questions that can be answered with a single word or short phrase) when seeking to understand human subjects' perceptions of an issue.	Use interviews to understand perceptions in detail. Collaborate with a social scientist.
Analysing social science data using quantitative analysis, thus losing a great deal of the most useful information.	Use qualitative analyses for social science. Collaborate with a social scientist.
Confusing the observed **rate** of an event (e.g. death due to predation) with the intrinsic **risk** of that event (the likelihood of dying due to predation in the absence of anti-predator defences). Measuring rate ignores the effects of any countermeasures in place, but risk is more difficult to measure. Examples are predation, infanticide, and competition.	Think carefully about what you can (and can't) measure.
Using simple patterns of behaviour to measure **preference** without testing whether the patterns differ from what we would expect from the availability of the resource.	Measure patterns of resource use relative to availability of the resource if you wish to understand preference.
Using inappropriate survey methods.	Understand survey design and the assumptions underlying the methods you use.

Box 14.1: (cont.)

Using remotely sensed data without good understanding of what it represents.	Calibrate remotely sensed data with real features on the ground (**ground-truthing**).
Using telemetry devices to assess habitat use without checking for biases in fix success due to environmental variables or animal behaviour.	Test telemetry devices and measure and account for any biases in fix success before drawing conclusions about habitat use.

FURTHER READING

Buckland ST, Plumptre AJ, Thomas L, Rexstad EA. 2010. Design and analysis of line transect surveys for primates. *International Journal of Primatology* 31: 833–847. https://doi.org/10.1007/s10764–010-9431-5. Reviews common errors in primate surveys, the assumptions underlying the standard method, and potential alternatives for when the standard approach won't work.

Rust NA, Abrams A, Challender DWS, Chapron G, Ghoddousi A, Glikman JA, Gowan CH, Hughes C, Rastogi A, Said A, Sutton A, Taylor N, Thomas S, Unnikrishnan H, Webber AS, Wordingham G, Hill CM. 2017. Quantity does not always mean quality: The importance of qualitative social science in conservation research. *Society and Natural Resources* 30: 1304–1310. https://doi.org/10.1080/08941920.2017.1333661. Explains when, why, and how qualitative methods should be used in conservation studies. Recommends collaboration between natural and social scientists.

Setchell JM, Fairet EFM, Shutt K, Waters S, Bell S. 2017. Biosocial conservation: Integrating biological and ethnographic methods to study human–primate interactions. *International Journal of Primatology* 35: 401–426. https://doi.org/10.1007/s10764–016-9938-5. Reviews three case studies combining natural and social science methods to yield insights that would not have been obtained using a single approach.

Shutt K., Setchell JM, Heistermann M. 2012. Non-invasive monitoring of physiological stress in the Western lowland gorilla (*Gorilla gorilla gorilla*): Validation of a fecal glucocorticoid assay and methods for practical application in the field. *General and Comparative Endocrinology* 179: 167–177. https://doi.org/10.1016/j.ygcen.2012.08.008. An example of a detailed biological and immunological validation of a non-invasive assay for glucocorticoids.

Städele V, Vigilant L. 2016. Strategies for determining kinship in wild populations using genetic data. *Ecology and Evolution* 6: 6107–6120. https://doi.org/10.1002/ece3.2346. Explains why it is difficult to assess kinship among members of wild animal populations without detailed multigenerational pedigrees and outlines ways to address this.

14.8 FURTHER READING

Martin P, Bateson PPG. 2007. *Measuring Behaviour: An Introductory Guide*. 3rd edn. Cambridge: Cambridge University Press. Excellent advice for measuring behaviour.

Methods in Ecology and Evolution. A journal of the British Ecological Society that *publishes* new methods in ecology and evolution.

Setchell JM, Curtis DJ. 2011. *Field and Laboratory Methods in Primatology: A Practical Guide*. 2nd edn. Cambridge: Cambridge University Press. Technical and practical aspects of field and laboratory methods, including remote sensing, GPS and radio-tracking, dietary ecology, and non-invasive genetics and endocrinology.

Wyatt TD. 2015. The search for human pheromones: The lost decades and the necessity of returning to first principles. *Proceedings of the Royal Society B* 282: 20142994. https://doi.org/10.1098/rspb.2014.2994. A cautionary tale of wasted scientific effort as a result of researchers not checking the evidence behind claims made in the literature.

15

Planning Data Analysis

Our study design (Chapter 12) dictates the statistical analyses we need. It is very common to dive straight into collecting data without a detailed plan for how we will test our predictions. This is partly because statistics can be intimidating. However, we must understand statistical analysis to understand the strengths, limitations, and potential biases of any research. Planning data analysis in detail, before we collect our data, helps to avoid later problems and the more we think about statistical analysis now, the better our study will be. Thinking carefully about what our data will look like, and how we will test our predictions statistically may well lead us to revise our predictions.

I begin this chapter with some basic concepts and the need to consult a statistician when we design a study. I introduce types of statistical test, then the critical distinction between true replicates and pseudoreplicates. Next, I cover how to determine what sort of analyses we need, based on key questions about our predictions, types of variable, and how many measures we have per subject. I review tests for differences and for associations, then describe more complicated statistical models. I review other useful statistical methods, and end with how to prepare a detailed analysis plan and how to collect data in the format we need for analysis.

15.1 GETTING STARTED

Statistical training is beyond the scope of this book, but the overview in this chapter should help you to understand the important connection between statistical analysis and study design. The material may all be familiar to you but if it's not, you have work to do.

As we saw in Chapter 5, we use inferential statistics to infer something about a statistical population based on a representative sample drawn from that population. Re-read Chapter 5 to remind yourself about null hypothesis statistical testing and what the p value does and doesn't tell you.

If you are not already familiar with a statistical software package, get trained, or train yourself, in how to use one. R, the most comprehensive, is free. There are many other options, some of which are expensive. Your institution may give you a licence to a package, or have a reduced rate for a licence, but that will only last while you are enrolled or employed, so a free package is better. Some packages have a steep initial learning curve, but once you are familiar with one package, it is easier to learn to use others.

All statistical tests are based on assumptions about the data you test, and you need to understand those assumptions. Statistical software performs analysis for us, but it only does what we ask it to do. It won't tell us if our thinking is flawed or if the test we run is inappropriate, so it is essential to understand how the tests work.

A good understanding of statistical analysis comes from analysing many different types of datasets. Examine the methods sections of papers to see how other researchers tested their predictions. Remember that they may have made errors that the reviewers did not pick up. Downloading datasets from published studies and reproducing the analyses is an excellent way to familiarise yourself with analysis.

Modern primatologists rarely do their statistical analyses unaided. Statistics is an active research area, and a skilled statistician who is familiar with your research area will be able to help you plan your analyses. Persevere with this. I recall a baffling meeting with a statistician in the early days of my PhD. I didn't go back, but I should have.

Websites and online advice fora are useful sources of information but are not always correct. Also beware of statistical rules of thumb; they may not be correct, either. Read the literature and consult experts.

Finally, although statistical analyses can get very complex, we don't always need extremely complicated statistical analyses to address our predictions.

15.2 TYPES OF STATISTICAL TEST

There's a broad range of types of statistical test, and the test you need depends on the prediction you want to test and the type of variable, as we'll see below. Before that, though, we need to understand the importance of the distribution of our data in our choice of statistical test. I also introduce one- and two-tailed tests, although one-tailed tests are rarely useful in primatology.

15.2.1 Parametric Tests, Non-parametric Tests, and Resampling

The type of statistical test we use depends on the distribution of our data.

Parametric tests make assumptions about the parameters of the population from which our data are drawn (Section 5.1). For example, they assume that the data in the population have a known distribution. This is often the normal distribution (Section 5.1). Parametric tests are powerful and quite robust to violations of the underlying assumptions.

Non-parametric tests make fewer assumptions about the probability distribution of the data. They are not, however, free of assumptions, so we still need to check that our data meet the assumptions of the test. Non-parametric tests work better for small sample sizes than parametric tests do, but can have lower statistical power, so we run a higher risk of a false negative (Section 5.6). In other words, non-parametric tests are more **conservative** than parametric tests. They are also quite intuitive. Non-parametric tests are useful if the median reflects the distribution better than the mean does, if you have a very small sample size, ordinal data (Section 5.2) or your data include outliers (Box 5.1).

Parametric models produce more accurate and precise estimates of model parameters than non-parametric models, because they have greater statistical power (Section 5.6). Where data do not meet the assumptions of parametric tests, we can often apply a simple mathematical function to each observation (a **transformation**) and use these transformed numbers in the statistical test. Common transformations include taking the logarithm or the square root of each observation. Moreover, many parametric models are fairly robust to violations of the assumptions.

Other alternatives for data that don't fit the assumptions of parametric tests include **resampling methods**, which avoid the problem of assumptions about the underlying distribution. For example, **bootstrapping** involves resampling the data repeatedly, with replacement (i.e. all values in the sample have an equal probability of being selected one or more times) and calculating the test statistic each time to obtain a distribution of values for the test statistic with which to compare your observed value. **Randomisation** or **permutation testing** involves shuffling observations among groups and recalculating the test statistic, then using the distribution of that statistic to determine the probability of the observed value.

15.2.2 One- and Two-Tailed Tests

Statistical tests can be **one-tailed** or **two-tailed**. The tails refer to the area of the distribution of the test statistic where values fall if a test is statistically significant. In a one-tailed test, the hypothesised relationship can only be in one direction. Two-tailed tests allow the relationship to be in either direction. One-tailed tests are rarely relevant to primatology, because we can rarely rule out all interest in an effect in the other direction.

15.3 INDEPENDENT REPLICATES AND THE PERILS OF PSEUDOREPLICATION

It can be tempting to count multiple samples from, or measurements of, the same individual as separate data points, because we want to obtain a larger sample size, or even simply an adequate sample size for analysis. However, as we saw in Section 12.4.2, factors that influence one measurement of an individual also influence other measurements of that same individual, so the measurements will be closely related to one another, or dependent. Treating such **pseudoreplicates** as statistically independent replicates (**true replicates**) artificially inflates our sample size, leading to a high risk of false positive results (rejecting the null hypothesis when it is true, Section 5.6), and erroneous conclusions. The sample size is the number of cases we study, and not the number of data points we obtain.

As an example, the body mass of a primate on one day is closely related to its body mass the day before. If we want to test whether female mandrills are lighter than male mandrills, measuring the body mass of a single female and a single male once a day for 10 days gives us two statistically independent replicates: one female and one male. It does not give us a sample size of 10 vs. 10, although it does improve our understanding of within-individual variation in body mass.

A common case of pseudoreplication in behavioural studies occurs when each subject appears in multiple **dyads**, resulting in a dyadic interaction matrix. We cannot treat dyads as independent data points. One option is to test the relationship between two such matrices (e.g. between dyadic interaction frequency in one matrix and dyadic relatedness in another) using permutation tests.

A further common error is to use the same stimulus for all subjects in a playback experiment. Doing so measures the reaction to a *specific* call, not to a *class* of call, because we only have one replicate for the call.

Avoiding, or accounting for, pseudoreplication requires thought and careful experimental design. We must be open about possible sources of non-independence in our data and conservative about the conclusions we draw. A good understanding of our study animals helps with this.

A simple solution to pseudoreplication is to average the data points for each case, weighting these values by the number of data points per case, if this varies. However, doing so removes information on within-individual variation and wastes a great deal of data.

As we'll see below, some statistical tests explicitly account for dependence in the data. If a variable we are interested in, or a confound we wish to account for, varies between repeated measures of the same subject, this is a useful within-subject design (Section 12.4.2), provided we design our study carefully to account for dependence between the measures.

Finally, sequential observations on the same subject may also not be independent due to **temporal autocorrelation**. For example, an animal's activity at one time-point may not be independent of its activity at the next time-point. We can also run into problems in spatial analyses, when observations made close to one another may be more likely to be similar than those made at a greater distance (**spatial autocorrelation**). Finally, closely related species are more likely to be similar than more distantly related species are. In other words, shared evolutionary history leads to **phylogenetic correlation**. Specialised statistics account for each of these situations.

15.4 CHOOSING THE RIGHT TEST

To choose the right statistical test, we need to know the answers to five questions:

1. *What type of prediction do I want to test?* As we saw in Chapter 12, we can predict differences between groups, differences within subjects, or associations between variables.
2. *Is my outcome variable categorical, ordinal, or continuous (Section 5.2)?*

3.   *If my outcome variable is continuous, is its distribution non-parametric or parametric (Section 15.2)?*

4.   *How many predictor variables will I have?* With one predictor variable, tests are relatively simple. With more than one predictor variable, they become more complicated.

5.   *How many measures will I have per subject?* If we have one measure per subject, then our data are independent. If we have more than one measure per subject, they are dependent.

With the answers to these questions in hand, you can read the next four sections which go through the tests for each type of prediction. We begin with relatively simple tests, and then move on to more complicated study designs with multiple predictor variables and multiple dependencies in the data.

15.5 TESTING FOR DIFFERENCES BETWEEN GROUPS

If we predict differences *between* groups (Section 12.4.1), with a single predictor variable, and a single measure of the outcome variable for each subject (i.e. our data are independent), the test we need depends on whether the outcome variable is categorical, ordinal, or continuous (Section 5.2), and whether the predictor variable is split into two groups or more than two groups (Table 15.1).

With a categorical outcome variable, and a predictor variable with two or more groups, we can use **Fisher's exact tests** (for small samples) or **chi-squared tests** (for larger samples). With an ordinal outcome variable and a predictor variable with two groups, we can use a **Mann–Whitney U test**; with a predictor variable with more than two groups, we can use a **Kruskal–Wallis test**. These non-parametric tests also apply to continuous outcome variables that do not meet the assumptions of parametric tests.

With a continuous outcome variable that meets the assumptions of parametric tests, and a predictor variable with two groups, we can use an **independent samples *t*-test**; with a predictor variable with more than two groups we can use analysis of variance (**ANOVA**).

An independent samples *t*-test is a simple case of a one-way ANOVA, but they are often taught separately.

If we have more than two groups of subjects, and wish to compare each group with the others, we can't just run lots of separate paired comparisons because running multiple statistical tests increases the

Table 15.1 *Statistical tests for differences between groups.*

Outcome variable	Predictor variable	Null hypothesis	Test	Example
Categorical	Two or more groups	There is no difference in the proportion of observations in different groups	Fisher's exact test (for small samples) or chi-squared test of independence (for large samples)	Compare the numbers of male and female offspring born to high- vs. low-ranking females
Ordinal or continuous but non-parametric	Two groups	A randomly selected value from one group is equally likely to be less than or greater than a randomly selected value from a second group	Mann–Whitney U test (also called Wilcoxon rank-sum test)	Compare the dominance ranks of females and males in a group
	More than two groups	The mean ranks of the groups are the same	Kruskal–Wallis test	Compare the energy content of different foods
Continuous, parametric	Two groups	There is no difference in the mean value of the outcome variable between groups	Independent samples t-test	Compare height in tree in two species
	More than two groups	There is no difference in the mean values of the outcome variable across groups	One-way analysis of variance (ANOVA)	Compare time spent digging holes across age classes

probability that we will find a false positive (a statistically significant result when there is no underlying effect, Section 5.6). To avoid this **family-wise error**, we first test for an effect across all the groups, then, if we find a significant effect, follow this with *post hoc* **tests** to compare all the combinations of group and determine where the significant difference lies.

We can extend ANOVA to test the influence of multiple predictor variables and how they interact on an outcome variable using a factorial ANOVA. For example, we can test the influence of sex and season on time spent feeding using a two-way ANOVA. Here the null hypothesis is that the mean time spent feeding, split by sex and season, does not differ, and that the two factors do not interact to influence the time spent feeding. In this analysis, sex and season are **main effects**, and the combined effect of the two is an interaction (Section 12.6).

Analysis of covariance (**ANCOVA**) extends ANOVA to include one or more continuous predictors. We use it to account for the effect of a covariate, not to test its effect on the outcome variable. In other words, the covariate is not the predictor of interest. For example, we might need to account for the effect of age as a covariate when testing the influence of sex on time spent travelling (if we think the influence of age is linear).

15.6 TESTING FOR DIFFERENCES WITHIN SUBJECTS

If we predict differences *within* subjects (Section 12.4.2), with more than one measure of each subject, our data are dependent and the test we use depends on the type of outcome variable (ordinal or continuous) and the number of times we measured each subject (twice or more than twice) (Table 15.2).

With an ordinal outcome variable, or one that does not conform to the assumptions of parametric tests, and two measures for each subject, we can use a **Wilcoxon signed-rank test**; with more than two measures for each subject, we can use a **Friedman's ANOVA**. With a continuous outcome variable that meets the assumptions of parametric tests, and two measures for each subject, we can use a **paired *t*-test**; with more than two measures for each subject, we can use a **repeated measures ANOVA**. Datasets with dependent data in primatology often fail to meet the requirements of a repeated measures ANOVA – we'll look at potential solutions to this later.

Table 15.2 *Statistical tests for differences within subjects.*

Outcome variable	Number of measures for each subject	Null hypothesis	Test	Example
Ordinal or continuous, non-parametric	Two	The median difference between pairs is 0	Wilcoxon signed-rank test	Compare dominance ranks of the same animals in two seasons
	More than two	There is no difference in the rankings of the outcome variable across groups of matched data	Friedman's ANOVA	Compare scores for the same individuals in multiple cognitive tasks
Continuous, parametric	Two	There is no difference between the means of the outcome variable in two groups of paired data	Paired t-test	Compare time individuals spend in proximity with kin and non-kin
	More than two	There is no difference between the means of the outcome variable in two or more groups of matched data	Repeated measures ANOVA	Compare rates of aggression received by females across stages of the reproductive cycle

15.7 TESTING FOR ASSOCIATIONS BETWEEN QUANTITATIVE VARIABLES

If we predict an association between quantitative (i.e. numerical) variables, the appropriate analysis depends on whether we hypothesise a simple association between variables, or whether we want to assess how change in one or more predictor variables influences change in an outcome variable (Table 15.3).

15.7.1 Testing Simple Associations

We test for simple associations between variables using correlation statistics. Correlation coefficients measure the strength of a relationship and range from −1 to +1. The sign of a correlation indicates the direction of the association between the two variables. If the sign is positive, then one variable tends to increase as the other increases. If the sign is negative, then one variable tends to increase as the other decreases. A indicates no relationship between the variables and 1 indicates a perfect relationship, where all data points fall exactly on a straight line.

Table 15.3 *Statistical tests for associations between quantitative variables.*

Prediction	Null hypothesis	Distribution of variables	Statistical test
Simple association between variables x and y	Variables x and y are not associated	Non-parametric	Kendal or Spearman correlation
		Parametric	Pearson correlation
Simple association between variables x and y, accounting for a third variable z	Variables x and y are not associated when we account for the effect of variable z	Non-parametric	Partial Kendal or Spearman correlation
		Parametric	Partial Pearson correlation
Change in one or more predictor variables (x) influences change in an outcome variable (y)	Change in variable x does not influence change in variable y	See Section 15.10	Regression analysis

Where both variables are continuous, and normally distributed, we can use a parametric **Pearson correlation** to examine the strength of the linear relationship between them. For example, we might predict a linear relationship between time spent grooming and time spent in close proximity to other individuals.

Where data do not meet the assumptions of normality, we can use a non-parametric **Kendall rank correlation** or a **Spearman correlation**. These require only ranked data. For example, we might predict a relationship between dominance rank and food intake. Kendall rank correlation works better for small samples than the Spearman correlation.

Partial correlations measure the relationship between two variables, accounting for the effect that a third variable has on one of those variables. For example, we might predict a relationship between dominance rank and number of offspring in females, accounting for age.

All correlation tests assume that cases are independent.

15.7.2 Modelling Relationships between a Predictor Variable and an Outcome Variable

If we want to assess how change in one or more variables *predicts* change in the outcome variable, we use **regression analysis** to describe the relationship between the variables using a mathematical equation (a statistical model). This model allows us to estimate the strength and direction of the association between the predictor and outcome variables, test the null hypothesis that variation in a predictor variable does not cause some of the variation in the outcome variable, and predict the value of the outcome variable based on a predictor variable.

Simple linear regression models how a single predictor variable, such as relatedness, predicts the value of an outcome variable, such as grooming received. The result is an equation that describes a straight line. In other words, if we plan to use a linear model to test an association, the predicted relationship should be a straight line.

Like correlation, simple linear regression assesses the linear relationship between two variables. However, unlike correlation, linear regression distinguishes between the predictor and the outcome variables and predicts the value of the outcome variable based on that of the predictor variable.

We can also test the fit of a curved relationship using **polynomial regression**. For example, we might predict a U-shaped relationship between male dominance rank and age in sexually dimorphic species where males compete physically over access to females, such that both young and old males are lower-ranking than males of prime age.

If our outcome variable is categorical and our predictor is continuous or categorical, we use **logistic regression**. For example, we might model the influence of prenatal maternal hormone levels on the sex of the offspring. Logistic regression can have a binary (in this case male vs. female) or multinomial outcome (one with more than two categories).

15.8 MORE COMPLICATED STATISTICAL MODELS

Most primatological datasets are complicated by both multiple predictor variables and multiple dependencies in the data, meaning that we need more complex models than those we have reviewed so far. These include multiple regression, generalised linear mixed models, and model selection. You're very likely to need expert statistical advice if you plan to use these models.

15.8.1 Multiple Regression

If we have enough data to do so, combining the effects of multiple predictor variables in a single model is better than using multiple independent tests of single predictor variables, for several reasons. First, using a single analysis allows us to assess how well we can predict the outcome using the entire set of predictors. Second, assessing their effects simultaneously allows us to assess the unique effect each predictor variable has on the outcome variable, accounting for the effect of the other predictors. Third, we can investigate interactions between predictors (Section 12.6). Fourth, multiple regression helps to avoid problems of multiple testing (Section 5.6).

We can extend simple regression to test how change in a combination of two or more predictor variables predicts change in the outcome variable using **multiple regression**. For example, we might predict that both dominance rank and relatedness influence grooming received, or we might account for the effect of a confounding variable (Section 5.3). We can also model interactions between predictor variables. For example, we might predict that the effect of relatedness on grooming received depends on the sex of the recipient. If we include an interaction term in a model, we must also include the independent effects of the variables in the interaction.

Regression analyses with one outcome variable and multiple predictor variables are **multivariable analyses**. This is not the same as **multivariate analyses**, which are statistical models that have two or more outcome variables, although this is a common mistake in terminology.

We need a sample size considerably greater than the number of predictor variables to avoid fitting a model that explains the data but doesn't generalise (**overfitting**). To help with this, we need to think carefully about the models, based on our understanding of the system we're studying, and use theory to narrow the range of likely predictors.

We can include categorical predictor variables in regression models by creating new variables to represent the categorical variable using only 0 and 1 (**dummy coding**). Statistical software packages will do this for you, but you need to know what they are doing to interpret the results. This is the equivalent of an ANOVA (Table 15.1), although statistical software reports different information in the output. Try running the two tests on the same data and compare the results.

It's common practice to divide one variable by another to produce a ratio variable, to remove the effect of the denominator from the analysis. For example, we might be tempted to divide brain mass by body mass, male mass by female mass or calculate the ratio of the length of the 2nd and 4th digits (2D:4D ratio). However, taking a ratio makes assumptions about the relationship between the two variables composing it, which may not be valid. As a result, using ratios can bias our inference or even lead to erroneous conclusions. To avoid this, we should include both variables in the model, rather than calculating a ratio. Similar arguments apply to using the differences between the observed values and the values predicted by the model (the **residuals**) from linear regression in further analysis. Again, we should include both variables in a multiple regression, rather than using the residuals of one variable regressed on the other.

We can further generalise linear regression to encompass outcome variables that are not normally distributed – the **generalised linear model**.

Box 15.1 provides a list of questions to ask yourself when putting together a regression model.

15.8.2 Generalised Linear Mixed Models

Where we have a continuous outcome variable, and measure each subject more than twice, datasets in primatology often have missing

Box 15.1: Data analysis plan for regression analyses

In null hypothesis significance testing (Chapter 5) using regression models, you wish to test whether a model that includes both your test predictors and your control predictors explains your data better than a null model that includes only the control predictors. This set of questions will help you to translate your verbal predictions into a statistical model. Write one of these for each prediction.

What Prediction Are You Testing?
- Write your prediction as I *predict that an increase in x will be associated with an increase/decrease in y.*

What Outcome Variable Will You Measure?
- The type of outcome variable determines the type of regression model you need.
- This is y in your prediction statement.
- If your outcome variable is continuous, you can use linear regression.
- If it is binary, you need logistic regression.
- If it is count data, you need Poisson regression.

What Are the Test Predictors?
- This is x in your prediction statement.
- You may have multiple predictors.
- These are the main effects.

What Are the Control Predictors?
- These are variables that you do not have specific predictions for, but which have effects on the outcome variable, which might prevent you from observing your predicted effect (i.e. confounds).
- For example, you might include the group, season, or phylogenetic effects.

What Are the Potential Interaction Effects?
- Interactions occur where the relationship between one variable and the outcome variable depends on the value of another variable.
- They can be between test predictors, test and control predictors, and between control predictors.
- If you include interaction effects in your model, remember to include the associated main effects.

Box 15.1: (cont.)

Are Any Relationships Likely to Be Non-Linear?
- Non-linear relationships occur where change in one variable is not directly proportional to that in the other.
- This has implications for the type of model you need.

Are the Predictors Random Effects or Fixed Effects?
- Fixed effects are where you can measure all values of the predictor. For example, you can measure all levels of *season* or *sex*.
- Random effects are where the values you can measure are a sample of a larger population. *Subject ID*, *group*, and *year* are common examples.
- This is a complex area and you'll probably need expert statistical guidance.

If you know what you're doing, you can now describe your model as an equation. If you don't, take your answers to these questions to a statistician for help.

---

data due to difficulties with sampling. This fails to meet the requirements of a repeated measures ANOVA (Table 15.2). Fortunately, **generalised linear mixed models** (GLMMs) are more flexible and therefore extremely useful. GLMMs allow us to account for and investigate both within- and between-individual levels of variation and allow various distributions of the outcome variable. They also allow hierarchical levels of organisation. For example if we have multiple measurements for individuals in different groups, we need to account for both ID and group in our analysis. GLMMs are very popular in primatology but are challenging to implement correctly. You are likely to need advice from experts.

15.8.3 Model Selection

Stepwise **model selection**, in which we fit a regression model using an automated procedure to select variables based on specific criteria (including forward selection, backward elimination, and bidirectional elimination), is a popular approach to modelling multiple predictors, and some textbooks recommend it. However, it is highly problematic for several reasons. First, stepwise selection performs poorly, in that parameter estimates are biased, it doesn't necessarily identify the *best* model, and different selection methods yield different answers. Second, it focusses attention on a single model, without examining other models with similar support. Third, it applies methods intended to test a single hypothesis multiple times, leading to inflated risk of false positives (Section 5.6). In other words, stepwise analysis is an example of multiple testing (Section 3.5). It is best to avoid this method.

Information-theoretic and Bayesian approaches to model selection and multi-model inference address the problem of multiple predictor variables by comparing the likelihood that one or more models (hypotheses) explain the data. These approaches are based on a different paradigm to null hypothesis statistical testing. As we saw in Chapter 5, null hypothesis testing tells us how likely our observed result is, given that the null hypothesis is true. In contrast, information-theoretic and Bayesian approaches to model selection measure the relative support for a set of candidate models, given the data. This is often closer to the question we really want to ask.

The distinction between exploratory and confirmatory analyses (Section 3.5) is crucial when using an information-theoretic or Bayesian approach to multi-model inference. If we begin with a carefully defined set of plausible alternative hypotheses (**candidate models**), derived from theory and previous work, then collect data and evaluate the support for each of them, we can use model selection in confirmatory analyses (i.e. hypothesis testing). However, if we use model selection to explore our data and find the best model(s) to explain it, which is more common, this is exploratory analysis (i.e. hypothesis generating) and we can use our findings to *propose* hypotheses, but we need an independent dataset to test those hypotheses with confirmatory analysis.

Model selection using information-theoretic and Bayesian approaches has the great advantage of not using arbitrary cut-offs for statistical significance and avoiding problems with multiple testing (Section 3.5). However, these approaches are not immune to the problem of arbitrary cut-offs. For example, some people compare models using Bayes factors with arbitrary, but informed, criteria for low/medium/high support of one model over another. We should not use arbitrary rules (such as an arbitrary value for change in information criteria) without thinking about why we are doing so.

A further point to bear in mind is that we should not combine information-theoretic model selection with null hypothesis statistical testing by selecting the best of many models, and then testing it using a null hypothesis statistical test. This practice involves the same multiple testing issue as with stepwise regression. We must also examine the best model(s) carefully, because it may be that none of the models have much inferential value (model selection doesn't tell us this). Finally, we must also compare models fitted to exactly the same observations. (Automated routines in statistical software often drop incomplete cases unless you instruct them not to.)

15.9 OTHER USEFUL STATISTICAL METHODS

Phew. We can get back to simpler tests. Some of the simplest statistical tests test the null hypothesis that there is no difference between observed frequencies in our data and those expected by chance. For example, we might compare the number of males and females born in a population with the expected sex ratio of 1:1. To do this, we use an exact test for goodness of fit for small samples or a chi-squared test of goodness of fit for large samples.

Other statistical methods are useful under specific circumstances. For example, we often plan to measure several, or even many, variables that may be correlated with one another. In such cases, we can use multivariate methods including **factor analysis** and **principal components analysis** to identify clusters of variables. These methods help us to understand the structure of a dataset or reduce a dataset to a manageable size, retaining as much information as possible. For example, we might test primates on a range of cognitive tasks and use factor analysis to identify clusters reflecting different cognitive abilities, or we might measure many different aspects of an animal's morphology and use principal components analysis to reduce these to a more manageable number of variables describing morphology.

Discriminant function analysis determines which variables discriminate between two or more predetermined groups. For example, we might measure many different variables relating to vocalisations and use discriminant function analysis to determine which variables best discriminate the sex of the caller, or we might measure many different characteristics of fruit and use discriminant function analysis to determine which variables best discriminate whether the fruit are eaten by primates. We can assess how accurately the model classifies the data by excluding one observation, using the remaining observations to generate a discriminant function, then using that function to classify the excluded observation (leave-one-out **cross-validation**).

Circular statistics are useful when the distribution of data is circular and high values are adjacent to low values. For example, in compass bearings, 0° equals 360°. Taking a sample mean isn't very useful to describe such data. For example, the mean of a compass bearing of 350° and 10° is 180°, but if you look at a compass, you'll see that 0° is a more appropriate mid-point. Circular statistics account for this. For example, the Rayleigh test tests for clustering in circular data. Other examples of circular variables are time of day, day of the month, and month of the year. If we use only part of the distribution, we don't need circular statistics, but if we use all of it, we do.

Survival analysis is useful when we are interested in factors affecting the time elapsed before an event occurs. For example, we might model the effect of sex on how long an animal lives or of high vs. low rank on the interval between two births for a female. Unlike other models, survival analysis accounts for *censored* data, when information about a case's survival time is incomplete. For example, it allows us to include animals who are still alive at the end of a study of lifespan, or who have not yet given birth again if we're interested in inter-birth intervals.

15.10 PREPARING A DETAILED ANALYSIS PLAN

Writing a detailed analysis plan is an essential part of preparation for a project. For confirmatory (hypothesis-testing) research, an analysis plan ensures that we collect data that we can use to test our predictions and protects us from the problems of data mining and *post hoc* theorising (Section 3.5). Planning our analyses in detail before we start collecting data can seem tricky at first, but with some thought it greatly improves the science.

Our plan should include checking that our data conform to the assumptions of the statistical tests we plan to use. For some tests, like Pearson correlation, the variables themselves should be normally distributed. For models based on regression (including *t*-tests, ANOVA, and linear regression), the residuals must be normally distributed. Moreover, the distribution of the residuals around the regression line should be equal (**homogeneity of variance** or **homoscedasticity**). The outcome and predictor variables don't need to be normally distributed (this is a common misconception), but it is easier to interpret the results if they are. Statistical packages provide diagnostic plots and tests of these assumptions, so practice interpreting these.

Strongly correlated (**collinear**) predictor variables cause trouble in regression models. Where there is an approximate linear relationship between two or more predictor variables (i.e. they are not independent), this can cause problems with model fitting and interpretation and the results for individual predictors may be inaccurate. For example, the number of animals in a group is strongly related to the number of any subset of those animals. In other words, the variables are **redundant**. If we include the total number of animals and each of the subsets, we have perfect multicollinearity, where one variable is a linear combination of the others. These are simple examples; others may be less obvious.

We can't know in advance whether our data meet the assumptions of a particular test. However, we can specify what the assumptions are and plan how we will test whether our data fit them and what we will do if they do not. This plan should include the transformations we will test and specify the alternative non-parametric or randomisation tests we will use if necessary. In other words, our analysis plan will include a decision tree, or flow chart, with defined decision points and outcomes.

Once you have a detailed analysis plan, you can include it in a preregistered project plan (more on this in Chapter 17).

15.11 COLLECTING DATA IN THE FORMAT NEEDED FOR ANALYSIS

Set up the spreadsheet that you will use in your analyses before you collect data. Think ahead and set up your spreadsheet in a format that is appropriate for transfer to statistical software for analysis. Keep data you will analyse together in a single worksheet.

Datasets are made up of columns and rows. The columns are variables. A variable contains all the values that measure the same underlying attribute (e.g. ID, date, mass, temperature, duration) across the cases you measured (e.g. across individual animals, samples or transects). Give each column a unique, brief, and informative heading in the first row of your spreadsheet. These are your variable names.

Rows are your observations. An observation contains all the values measured for the same case across attributes. You shouldn't have more than one row per case, and you shouldn't have more than one case per row.

Don't split a variable into different columns for different groups (e.g. wet season values in one column, dry season values in another column). Instead place all the values for a variable in a single column and use a second column (in this case season) to describe the groups. The only exception to this is if you intend to use the tests for matched data in

Table 15.2. With large datasets, you will probably need multiple grouping variables. Use codes, not colours, to differentiate your data, so that you can easily import them into a statistical package. Don't use extra lines for notes (place these in a separate sheet) and don't merge cells. Check how you should code missing data for the statistical package you plan to use. For example, don't use 0 for missing data, because it is a value.

Make your spreadsheet self-explanatory, such that another researcher would understand it without needing your help.

15.12 CHAPTER SUMMARY

In this chapter, we've seen that:

- A detailed data analysis plan helps to determine which data we need to collect.
- Without careful thought at this stage, we risk spending months or years collecting data that cannot answer the question(s) we wish to answer.
- Primatological datasets are often complex and most primatologists consult or collaborate with expert statisticians for data analysis.
- Our sample size is the number of independent cases we measure, not the number of measurements we obtain.
- The type of test we need depends on the prediction, the type of variables we have, and the number of measurements we have for each subject.
- Simple tests include differences between groups, differences within subjects and associations between two quantitative variables.
- Datasets with multiple predictor variables and multiple dependencies in the data need more complex models.
- Writing a detailed analysis plan protects us from the problems of data mining and *post hoc* theorising.
- We should collect data in the format that we will analyse it.

In the next chapter, we look at how we select an appropriate sample.

15.13 FURTHER READING

Barnard C, Gilbert F, McGregor P. 2017. *Asking Questions in Biology: A Guide to Hypothesis Testing, Experimental Design and Presentation in Practical Work and Research Projects*. 5th edn. Harlow, Essex: Benjamin Cummings. Chapter 3 covers statistical analysis and hypothesis testing. Also includes quick test-finders at the end of the book, linking statistical tests to study design. Recommends step-wise regression but don't follow this advice.

Beckerman A, Petchey O, Childs D. 2017. *Getting Started with R*. Oxford: Oxford University Press. An introductory guide to the standard software for statistical analyses and graphical presentation of data. Includes how to import, explore, plot, and analyse data.

Burnham KP, Anderson DR. 2002. *Model Selection and Multimodel Inference: A Practical Information-Theoretic Approach*. 2nd edn. New York: Springer-Verlag. An introduction to information-theoretic approaches to model selection and multi-model inference.

Darlington RB, Smulders TV. 2001. Problems with residual analysis. *Animal Behaviour* 62: 599–602. https://doi.org/10.1006/anbe.2001.1806. Explains sources of bias in residual analysis, and recommends multiple regression.

Field A, Hole G. 2003. *How to Design and Report Experiments*. London: Sage Publications Ltd. Part 2 covers analysing and interpreting data.

Field A, Miles J, Field Z. 2012. *Discovering Statistics Using R*. SAGE Publications Ltd. Includes the logic behind tests, how to do them, and how to report the results, with worked examples. Uses *theory* for what I term a *hypothesis* and *hypothesis* for what I term *predictions*.

Freckleton RP. 2002. On the misuse of residuals in ecology: regression of residuals vs. multiple regression. *Journal of Animal Ecology* 71: 542–545. https://doi.org/10.1046/j.1365-2656.2002.00618.x. Explains why we shouldn't treat the residuals from regression as data in further analysis.

Garamszegi LZ. 2011. Special Issue on 'Model selection, multimodel inference and information-theoretic approaches in behavioural ecology'. *Behavioral Ecology and Sociobiology* 65: 1–116. A collection of articles from authors with different viewpoints on how to analyse data with multiple predictors. Essential reading if you intend to use an information-theory approach.

Grueber CE, Nakagawa S, Laws RJ, Jamieson IG. 2011. Multimodel inference in ecology and evolution: Challenges and solutions. *Journal of Evolutionary Biology* 24: 699–711. https://doi.org/10.1111/j.1420-9101.2010.02210.x. An overview of multi-model inference, common challenges faced when using the information theoretic approach and potential solutions where they exist.

Harrison XA, Donaldson L, Correa-Cano ME, Evans J, Fisher DN, Goodwin C, Robinson BS, Hodgson DJ, Inger R. 2018. A brief introduction to mixed effects modelling and multi-model inference in ecology. *PeerJ* 6: e4794. doi:10.7717/peerj.4794. A best practice guide for the application of mixed effects models and model selection in ecological studies.

Janson C. 2012. Reconciling rigor and range: observations, experiments, and quasi-experiments in field primatology. *International Journal of Primatology* 33: 520–541. https://doi.org/10.1007/s10764-011-9550-7. Includes the use of random-effects models to account for repeated and uneven sampling of the same individuals.

Kronmal RA. 1993. Spurious correlation and the fallacy of the ratio standard revisited *Journal of the Royal Statistical Society. Series A (Statistics in Society)*. 156: 379–392. http://dx.doi.org/10.2307/2983064. Shows that using ratios in regression analyses can lead to incorrect or misleading inferences.

Kroodsma DE, Ryers BE, Goodale E, Johnson S, Liu W-C. 2001. Pseudoreplication in playback experiments, revisited a decade later. *Animal Behaviour* 61: 1029–1033. https://doi.org/10.1006/anbe.2000.1676. Examines the issue of pseudo-replication in playback experiments and the need for multiple stimuli to represent a class of stimuli.

Mangiafico SS. 2015. An R Companion for the Handbook of Biological Statistics, version 1.3.2. rcompanion.org/rcompanion/. Pdf version: rcompanion.org/documents/RCompanionBioStatistics.pdf. Notes on getting started in R, and R code for many of the examples given in McDonald's *Handbook of Biological Statistics* (below).

McDonald JH. 2014. *Handbook of Biological Statistics*. 3rd edn. Baltimore, MD: Sparky House Publishing. Available free online at http://biostathandbook.com/ and as a pdf. A very useful introduction to statistics, including links to spreadsheets to calculate simple tests.

McElreath, R. 2015. *Statistical Rethinking: A Bayesian Course with Examples in R and Stan*. Boca Raton, FL: CRC Press. An excellent, pragmatic introduction to the reasoning underlying Bayesian inference, from basic probability to generalised linear multilevel modelling. Recorded lectures available online.

McGregor PK. 2000. Playback experiments: Design and analysis. *Acta Ethologica* 3: 3–8. https://doi.org/10.1007/s102110000023. Examines issues in the design of playback experiments, including pseudoreplication.

Mundry R, Nunn R. 2009. Stepwise model fitting and statistical inference: Turning noise into signal pollution. *American Naturalist* 173: 119–123. https://doi.org/10.1086/593303. Shows why we should not use significance tests based on stepwise procedures.

Mundry R, Sommer C. 2007. Discriminant function analysis with non-independent data: Consequences and an alternative. *Animal Behaviour* 74: 965–976. https://doi.org/10.1016/j.anbehav.2006.12.028. Explains that including non-independent data in a discriminant function analysis yields incorrect results, and provides a permutation-based test that copes with such datasets.

Nosek BA, Ebersole CR, DeHaven AC, Mellor DT. 2018. The preregistration revolution. *Proceedings of the National Academy of Sciences of the United States of America* 115: 2600–2606. https://doi.org/10.1073/pnas.1708274114. Includes the idea of a decision tree defining a sequence of tests and decision rules at each stage of analysis.

Ruxton GB, Neuhäuser M. 2010. When should we use one-tailed hypothesis testing? *Methods in Ecology and Evolution* 1: 114–117. https://doi.org/10.1111/j.2041-210X.2010.00014.x. Highlights how one-tailed tests are often used without clear justification.

Stamp Dawkins M. 2007. *Observing Animal Behaviour: Design and Analysis of Quantitative Data*. Oxford: Oxford University Press. Chapter 4 addresses independence of data and matching each case to itself.

Waller B, Warmelink L, Liebal K, Micheletta, Slocombe KE. 2013. Pseudoreplication: A widespread problem in primate communication research. *Animal Behaviour* 86: 483–488. https://doi.org/10.1016/j.anbehav.2013.05.038. Highlights the problem of pseudoreplication in primate communication studies.

16

Sampling and Statistical Power

Good research design includes careful consideration of the number of independent replicates we need to test our predictions: the sample size. Some sampling decisions are beyond our control. For example, we may be limited by the number of specimens available, the animals we can observe, or the data we have at our disposal. Knowing in advance what we can and can't test with our data will save wasted effort.

This chapter covers sampling strategies, the importance of statistical power, how to determine whether you have the power to test for an effect, and statistical precision.

16.1 SAMPLING METHODS

As we saw in Section 5.1, we usually can't measure our study variables for the whole population, so instead we select and measure a sample from that population. If we wish to generalise the parameters we derive from our sample to the population, this sample must be representative of that population (i.e. it must be externally valid, Section 13.3). **Sampling bias** arises where the sample is not representative.

Sampling involves defining a population, then specifying a sampling scheme to select the individual animals, groups, plants, transects, specimens, and so on to measure, and how often to measure them if we take repeated measurements. A sampling strategy must be tailored to the aim, so there are no standard protocols. This applies to all the populations you sample, including, but not limited to, the animals you study. If you wish to measure phenology, for example, you also need a representative sample.

It is not always easy to define a population and therefore a sample which is representative of that population. There are no simple rules.

Instead, we should describe our sample carefully and be clear when we are generalising from that sample.

Sampling methods guard against bias, to ensure that the samples we select represent the population we wish to study. They include random, systematic, stratified, haphazard, and opportunistic sampling.

In **random sampling** each member of population has an equal chance of being chosen, and all individuals are chosen independently. We use a random number generator to select cases.

Random sampling is not always practical. For example, we often study members of the same primate group, so selection can be random within that group but is not random within the population. Moreover, we often select a convenient, semi-habituated, or large study group, rather than a random group.

Random sampling can lead to underrepresentation of some subjects or categories. A pragmatic solution to this is to begin with random sampling, then balance the resulting coverage and equalise representation.

Systematic sampling involves choosing a random starting point, then sampling at a fixed interval. It can be biased if the sampling interval matches a cyclical pattern in the population (e.g. daily rhythms or seasons).

When a population includes distinct categories, for example animals vary in sex, age class, or dominance rank, or habitat types vary across a study area, **stratified sampling** involves sampling each of these strata as an independent sub-population. Here, stratified sampling produces better estimates of population parameters than random sampling.

Haphazard sampling is an attempt at random sampling without using a random number generator. It is likely to suffer from sampling bias and thus be unrepresentative of the population.

Opportunistic (or **convenience**) **sampling** involves sampling what we see or can get our hands on. It is not representative of the population but can be useful for pilot studies.

16.2 STATISTICAL POWER

As we saw in Chapter 5, the statistical power of a null hypothesis statistical hypothesis test is the probability that it will reject the null hypothesis when the alternative hypothesis is true. In other words, power is the probability that a study will detect an effect when an effect exists. It is influenced by several factors, including:

- The effect size (Section 5.7). The larger the effect, the easier it is to detect.
- The sample size. The more data you have, the smaller the effect you can detect, because a larger sample size reduces the sampling error. Remember that the sample size is the number of true replicates, not pseudoreplicates (Section 15.3).
- The risk of a false positive (α) (Section 5.6).
- The precision of the response variable. Reducing the differences between the measured value and the true value (measurement error, Section 14.4) improves power.

Power is inversely related to the probability of concluding that there is no effect when in fact there is one (a false negative, Section 5.6). This probability is the β value of your test. In other words, insufficient power increases the risk of missing real effects. A great deal of published research is underpowered, and the results of such studies can be wrongly interpreted as evidence of no effect (Section 5.5). If we summarise work only in terms of significance of the effect, as researchers frequently do in literature reviews, then we commit this error. Worse, underpowered studies have very variable results, so are also more likely to give a surprising, significant, but false positive result.

You might be tempted to go for a very large sample size, if possible, to be sure of having sufficient power. However, overpowered studies are an inefficient use of resources, and can be unethical if they involve animals (Chapter 2). They can be as misleading as underpowered studies if not interpreted appropriately, because they can detect significant but trivial effects.

Once you understand the concept of statistical power, you can determine whether published studies are underpowered.

16.3 DETERMINING AN APPROPRIATE SAMPLE SIZE

We can use a **prospective power analysis** (or *a priori* **power analysis**) to determine the appropriate sample size for a study. We can't rely on intuition, advice from other researchers, or statistical rules of thumb. We must assess the sample size in previous studies critically; they may be underpowered.

Power analyses are easy for simple statistical models. They are very useful for grant applications and some funders require them. Things get more complicated if you plan to use more complex models and you'll probably need to consult a statistician. You may get a shock if the sample size required to test a prediction is unfeasibly large, but it's better to know this before you start.

The problematic part of a power analysis is estimating the effect size. Ideally, you should use a meta-analysis of previous research to do this (Section 11.8). Where that's not available, pool the data in the literature or use theory and what you know of your study system to estimate a plausible range of effect sizes. You can use data from a pilot study, but remember that pilot studies are underpowered, so the estimated effect size will be imprecise. There are conventions for small, medium, and large effect sizes, but these are not very helpful in the absence of the theoretical context. It is better to think for yourself and justify your choices based on what we already know. Your study should have sufficient power to detect the smallest effect size that you would consider biologically meaningful. If the sample size needed to detect an effect is unfeasible, as is often the case, then the conclusions you can draw based on your study will be limited.

Use your power analysis to balance the competing risks of false positives and false negatives (Section 5.6) for your study. Few authors consider this explicitly. Choosing a lower rate of false positives is often viewed as rigorous, but this is misguided as it can lead to an unacceptably high risk of failing to detect an existing relationship (a false negative). Consider what is appropriate for your study carefully. Power ≥ 0.8 is a generally accepted standard and gives you an 80% chance of detecting a real effect of the expected size, but don't just follow this standard without thinking.

We can improve our statistical power by reducing measurement error by using a within-subject study design (Section 12.4.2), good controls, and reliable measures (Section 14.3), and choosing the most powerful statistical test for our data (Section 15.2.1).

Journal reviewers may ask you for a retrospective power analysis, usually if your study findings are non-significant. These are usually meaningless (Box 16.1), so if this happens, explain politely to the editor why the analysis is not useful.

16.4 DETERMINING HOW PRECISE YOUR ESTIMATE WILL BE

Most studies aim to test a null hypothesis and estimate the effect size with confidence, although this may not be explicit. Like power, the

Box 16.1: What's wrong with *post hoc* power analyses

Researchers sometimes run power analyses retrospectively, usually only when their results are non-significant. Reviewers and editors might even require this. However, such analyses are typically a waste of time. The motivation is good, because researchers want to know whether their results truly indicate the absence of an effect or whether they may have failed to detect an effect. However, a power analysis does not answer this question, for several reasons:

1. Calculating power assumes that the null hypothesis is false, but a non-significant result does not tell you this (Chapter 5). In other words, you don't have enough information to answer the question of whether you have failed to detect an effect.
2. If you use the effect size (see Chapter 5) you observed in a power calculation, then you're assuming that the actual population effect size is the same as your observed effect. However, if your power is low, then you can't be confident in this.
3. If you use the p value you observed in a power calculation, then all you're doing with a power calculation is re-stating the p value, because the power is directly related to the p value.

It is tempting to run a prospective power analysis after your test, to determine how large a sample size you would have needed to detect a particular effect size (Section 5.7), or the effect size you could detect with your sample size (the *detectable effect size*). Unfortunately, these results cannot help you interpret your non-significant results, but they can be used to assess the sensitivity of future studies.

Instead of focussing on the p value and conducting *post hoc* power analyses, examine your confidence intervals (Section 5.8) to determine how confident you can be in your parameter estimates.

FURTHER READING

Hoenig JM, Heisey DM. 2001. The abuse of power: The pervasive fallacy of power calculations for data analysis. *The American Statistician* 55: 19–24. https://doi.org/10.1198/000313001300339897. Shows that *post hoc* power calculations are common but flawed.

precision of the answer to a question improves with sample size. However, a sample large enough to test your prediction statistically may not be large enough to provide a precise estimate of an effect.

Understanding the relationship between sample size and the margin of error underlies *precision for planning* or *accuracy in parameter estimation* approaches. These approaches are less well-known than power analysis, and are little used in primatology, but have the advantage of moving the focus away from the dichotomy of null hypothesis statistical testing and a focus on statistical significance towards parameter estimation and the level of precision we regard as acceptable.

16.5 CHAPTER SUMMARY

In this chapter, we've seen that:

- Research design includes careful consideration of the number of replicates we need to test our predictions.
- Knowing in advance what we can and can't test with our data saves wasted effort.
- We aim to measure a representative sample of a population.
- A great deal of research is underpowered, leading to increased risk of false negatives and false positives.
- We use prospective power analyses to make informed judgements about sample size, minimum detectable effect sizes, and the relative risk of false positives and false negatives.
- Careful study design improves power.
- *Post hoc* power analyses are usually meaningless.
- A sample large enough to test a prediction statistically may be too small to estimate the effect size precisely.

Box 16.2 addresses common problems with sampling and how to address them.

Box 16.2: Common problems with sampling and how to resolve them

Common problems with sampling include data-peeking (Section 3.5), and pseudo-replication (Section 15.3), as well as the following issues:

Box 16.2: (cont.)

Sampling Bias
The sample is not representative of the general population we wish
to answer questions about. Resolve this by either choosing a
representative sample or redefining the population you're
sampling. Be clear when you generalise beyond the population
sampled.

Sample Size Is Too Small to Test the Predictions Adequately
Conduct a power analysis to determine the sample size. If you
cannot achieve an appropriate sample size, revise your study aims
to focus on a question you can address with the sample sizes you
can obtain.

It is tempting to determine sample size based on the resources we
have available. Of course, logistics constrain sample size. However,
the sample must be sufficient to test our predictions appropriately.

Stating that a previous study used similar numbers is not a good
justification for the sample size because that study may be
underpowered.

Sample sizes can also be too large, although this is not a common
problem in primatology. Very large samples risk detecting trivial
effects and may be unethical.

Lack of Understanding of the Purpose of Sampling
Students often worry that they can't measure everything. However,
all studies are selective. The key is to sample appropriately.

Conflating the Alpha Level with Rigour
Reducing α to reduce the risk of a false positives (Section 5.6) can
lead to a high risk of false negatives. Consider the balance between
the risk of false positive and false negatives.

In the next chapter, we take a reality check and look at the
feasibility of our plans.

16.6 FURTHER READING

Altmann J. 1974. Observational study of behavior: Sampling methods. *Behaviour* 49:
227–267. https://doi.org/10.1163/156853974X00534. A paradigmatic example
of thinking through the strengths and weaknesses of different sampling
regimes.

Cumming G, Calin-Jageman R. 2012. *Understanding the New Statistics: Estimation, Open Science and Beyond*. New York: Routledge. Includes *precision for planning* or *accuracy in parameter estimation* approaches.

Ellis PD. 2010. *The Essential Guide to Effect Sizes: Statistical Power, Meta-Analysis, and the Interpretation of Research Results*. Cambridge: Cambridge University Press. Written for the social sciences, but just as useful for primatology. Clear, succinct, and jargon-free.

Field A, Hole G. 2003. *How to Design and Report Experiments*. London: Sage Publications Ltd. Covers power on pages 154–157.

Johnson PCD, Barry SJE, Ferguson HM, Müller P. 2015. Power analysis for generalized linear mixed models in ecology and evolution. *Methods in Ecology and Evolution* 6: 133–142. https://doi.org/10.1111/2041-210X.12306. A clear discussion of power, and a simulation-based power analysis method appropriate for generalised linear mixed models.

Kelley K, Maxwell SE. 2003. Sample size for multiple regression: Obtaining regression coefficients that are accurate, not simply significant. *Psychological Methods* 8: 305–321. Introduces *accuracy in parameter estimation* (AIPE).

Reinhart A. 2015. *Statistics Done Wrong: The Woefully Complete Guide*. San Francisco, CA: No Starch Press. Chapter 2 covers statistical power.

17

Checking Feasibility and Finalising Your Plans

Good research requires good planning, and practical constraints influence the research questions we can address. Developing a detailed research plan helps us to determine the feasibility of a study and anticipate the issues we will face during it.

In this chapter, I cover logistics, using pilot studies to test the feasibility of a project, making a timeline, assessing risk, budgeting, and creating a detailed project plan.

17.1 LOGISTICS

Logistics are all the practical elements of your project. This includes the permits you need (Section 2.2), planning travel, and shipping equipment, consumables, and samples, if needed. Think through the research supplies you need, and how you will obtain them. What can you obtain locally and what might you need to take with you if you're travelling? If you're working cross-nationally, are there restrictions on what you can import and export? Airlines have restrictions on shipping liquids, particularly ethanol, and potentially infectious material.

Get familiar with your equipment and practice using it before you begin your study. Consider exactly how you will record, store, and back-up data and how you will preserve samples. Automated recording can produce a great deal of data to back-up. If you intend to work away from home, plan where you will stay, and where and what you will eat. If you need research assistants, consider how you will recruit them, and the training they will need (Section 2.8). Think about what might happen and what you would do. For example, what would you do if your equipment fails?

Logistics also includes the availability of study specimens, animals, or data and how your plans fit in with those of your research site (Section 12.8). For example, you may need to fit in with other projects running at the same place and time. Sites may be fully committed and unable to accommodate you. Collections may close for maintenance or loan the samples you're interested in to another institution. Facilities have opening hours and daily husbandry schedules that you must work around. Zoo animals must be available for public viewing and members of the public may wish to know about your research. Plan for this, so that you can be courteous without disrupting your data collection. For example, you might put up a sign asking people not to disturb you because you are collecting data and saying that you are available for questions at specified times. Alternatively, you can provide information about your research in a poster. In the field, your study animals may be right there at your site, and already habituated, but this is not always the case. Some species are difficult, or even impossible, to habituate.

17.2 ASSESSING RISK AND PLANNING FOR EMERGENCIES

Take responsibility for your own safety during data collection, and that of anyone who works for you. Discuss risks with people who are familiar with the places you'll be working and ask for any safety guidelines and protocols. Your institution may require a formal risk assessment.

Consider what might cause harm to you or people working with you and plan reasonable precautions to control the risks you identify. Don't assume that good safety measures are in place in a working environment. Potential risks include hazardous chemicals, sharp tools and objects, radiation, freezing due to contact with liquid nitrogen, electrocution, and biohazards. Personal protective equipment includes gloves, safety glasses, lab coats, shields, and appropriate footwear. Laboratory safety equipment includes sharps and biohazard disposal containers. Learn how to use equipment safely and read the manual. We'll cover the risks involved in fieldwork in Chapter 20.

Consider what might constitute an emergency and how to respond to it. Agree on, record, and rehearse emergency plans with those you work with, including actions and who is responsible for what. Get travel insurance, if that's appropriate, and check that the coverage is suitable.

Safety includes safety from discrimination. Discuss potential research sites with a trusted advisor and request a code of conduct (Section 4.11).

17.3 CONDUCTING A PILOT STUDY

Use a pilot study to assess the practicalities of everything you need to do to collect your data. This will help you to anticipate what might go wrong with your project, troubleshoot, and make contingency plans. If you can't do a pilot study, treat the first few weeks of your study as a pilot phase, then stop and assess your progress.

During your pilot study, meet everyone whom you will work with to discuss and agree on expectations. Discuss your plans, feasibility, and logistics in detail. Test all your equipment under the conditions that you will use it in. Test, validate, and practice your methods, refining them until they are repeatable, consistent, and accurate (Chapter 14). Methods established for other species may not work for your study species, so test them carefully.

Check that you can obtain the data you need to test your predictions. Practice sampling and check that your methods work. Check that you will be able to obtain the samples you need, and in sufficient numbers. Check that you can store biological samples appropriately, and transport them, if necessary. Check that any subsequent laboratory analyses work.

Check that the data you collect will allow you to test your predictions. Enter your pilot data into your data analysis spreadsheet and compute the summary variables you need for your analysis. Check for any gaps. Use your pilot data to estimate your effect sizes for power analysis (Section 16.3).

Also use your pilot study to look out for new and exciting possibilities.

Reflect on the implications of your pilot study for your project and study design. Discuss your findings with colleagues and seek their feedback. If you discover that you can't do something that you planned to do at all, or that you won't be able to obtain enough samples to test your predictions, redesign your study. Don't pursue a study design that's flawed. This may be demoralising, but it is normal and is why we need pilot studies. It is much better to find out that you can't address your question at this point than to find out at the analysis stage, or when you try to publish your work.

Be flexible and creative. There are many questions to ask and your experiences during your pilot study can make your project much more exciting than your original plan.

17.4 MAKING A TIMELINE

A project schedule, or **timeline**, is an essential part of planning. Timelines are useful at all stages of a project, including preparation, analysis, and writing up, as well as data collection.

To make a timeline, begin with your aims, then break them down into smaller manageable tasks, until you have listed all the tasks you need to accomplish to complete the project. Estimate how long each task will take, based on your own previous work and discussion with experienced colleagues. Be realistic with the amount of time you need for each task, not aspirational. Both inexperienced and experienced researchers usually underestimate the time needed to accomplish a task, so take this into account. Consider optimistic and pessimistic estimates for each task, to model various scenarios that may occur and the implications for the progress of your project.

Include time for permits, feedback on proposals, plans, and reports, and for acting on that feedback. Include time to input data, if you'll need to do that. Include breaks, and time for life outside your project.

Project management software will create a project schedule (a **Gantt chart**) for you if you list your tasks with their start and finish dates, or you can make one for yourself in a spreadsheet. List your tasks on the vertical axis of your timeline, and time intervals on the horizontal axis, from the start date to the end date of the project. The appropriate time intervals depend on the project length, but biweekly or monthly intervals are often an appropriate level of detail. Plot when you will undertake each task as a bar on the chart, to map out what you will do and when you will do it. You may need to include multiple people.

Some tasks rely on the completion of other tasks. For example, you can't start data collection until you have all your permits, and you can't analyse samples until you have collected them. Other tasks can run concurrently – for example different types of data collection. However, you can't usually do two things at the same time, so include how you will divide your time between multiple tasks. A timeline that says you'll do everything, all the time, is not helpful.

Prioritise your tasks, splitting them into things you have to do for your project to succeed, and things you'd like to do if you have time, but that are not essential.

Use your timeline to help identify risks and key points during the project (**milestones**). Include contingencies to allow for unforeseen events (i.e. slippage). For example, animals may go missing in the wild, captive facilities or labs may close for maintenance, and research assistants may drop out. Have a plan B and a plan C. Ask yourself *What if ...?* questions, and plan for the worst-case scenario.

Use your timeline to create to-do lists for each stage of your project. Make the tasks specific, measurable, assignable, realistic, and time-related (Box 17.1). In other words, plan what you need to do, when and how. If you plan a long period of data collection, set interim goals so that you can monitor your progress as you go along and avoid procrastination. For example, split 12 months of data collection into smaller weekly goals. The same applies to analysis and writing.

Box 17.1 SMART goals

SMART goals are very useful in project planning and monitoring. The words assigned to the letters vary, but here's an example: SMART goals are:

Specific – Goals should be clearly defined. What exactly do you want to accomplish?

Measurable – You need to be able to track your progress. How will you measure it? How will you determine whether you have met the goal?

Achievable – Goals should be realistic. Are they attainable, given the resources at your disposal?

Relevant – Goals should be appropriate to the overall project aim. Are your goals related to what you want to accomplish?

Time-bound – Goals need a deadline. When will you meet the goal?

Write protocols for each task. Test them and revise them until they are sufficiently clear and detailed that if you handed them to someone else, they could conduct the same project. These will underpin the methods sections of your reports.

17.5 BUDGETING

Think through the details of your project plan and estimate how much they will cost. Include salaries, supplies, permits, travel for data collection, equipment, consumables, shipping samples if necessary, and conferences. Be realistic and do your best to anticipate all the costs associated with your project. An underestimated budget at the planning stage will compromise your ability to conduct your project. Include possible currency fluctuations. Budgeting is not easy if you're new to it; get example budgets for similar projects and seek advice.

17.6 WRITING AND PREREGISTERING A DETAILED PROJECT PLAN

At this stage, you should be able to write a detailed and feasible project plan, including your research question, case study, hypothesis, predictions, study design, variables, sample size, and data analysis plan (Box 17.2). Doing so clarifies your thoughts and ensures you can share your plan with other people to get feedback.

Consider pre-registering your study (Section 3.5). Preregistration forces you to prepare a detailed plan – including your hypotheses, sampling plan, variables, study design, and data analysis – before you collect your data. Doing this avoids many of the problems associated with multiple testing and shows that you developed your predictions from theory rather than fishing for interesting results in existing data (Section 3.5).

You can pre-register a study in a dedicated registry (e.g. the Open Science Framework, https://osf.io) or as part of the publication process. In the registered reports publication model, researchers submit their research plans to a journal for peer review before conducting the research. Reviewers assess the plans, and if they find the question interesting and important, and the study design appropriate to address the question, the journal accepts the study in principle. The researchers then conduct the study, following the accepted methods and submit their final report to the journal. A second stage of review focusses on how well the study conforms to the accepted plans. Studies are not evaluated on how exciting the outcome is perceived to be, avoiding many of the questionable research practices we examined in Chapter 3.

Preregistered research plans can be embargoed, alleviating concerns that someone might copy your ideas. Moreover, preregistration provides evidence of when you registered your plan, which should further allay such concerns.

The potential for peer review of research plans is a great opportunity for primatologists, providing a chance for feedback on study design prior to data collection and potentially avoiding a great deal of wasted effort and much heartache. Preregistration represents a cultural shift in science and is just beginning in primatology.

Box 17.2: Project outline

By this stage of a project, you should be able to fill in the following information. The more detail you consider at this stage, the better your project will be.

My research question is:	State the phenomenon or observation you want to explain (Chapter 9)
This is interesting because:	Summarise the justification for your research question, based on your review of the literature (Chapter 11)
My case study is:	State the study system you intend to use (Section 9.2)
My case study is appropriate because:	Summarise the justification for your choice of case study, based on your review of the literature (Chapter 11)
My hypothesis is/hypotheses are:	State the potential explanation(s) for the phenomenon or observation under consideration (Section 12.1)
My predictions are:	List the logical consequences of your hypothesis/es as *If ... then ...* statements (Section 12.2)
My study design is:	State whether your comparisons are between groups or within subjects (Sections 12.4–12.7)
To test the predictions, I will:	State whether you will use observation or manipulation (Chapter 13)
The variables I need to test my predictions are:	List exactly what you need to measure to test the predictions. Describe each variable that you will measure and how you will combine them (if relevant) Justify the proxies you choose (Section 12.3)
To measure these variables, I will:	State the methods you will use to measure each variable (Chapter 14)
To test my predictions, I will use:	Explain the data analysis you will use to test each prediction (Chapter 15)
The population I want to sample is:	State the population you will sample (Chapter 16)
To obtain a representative sample of this population, I will:	Explain the type of sampling you will use (Section 16.1)
To test my predictions, I need a sample size of:	State the sample size you will need (Chapter 16)

17.7 CHAPTER SUMMARY

In this chapter, we've seen that:

- A detailed research plan helps to determine the feasibility of a study and anticipate problems.
- Logistics include equipment, consumables, travel, accommodation, assistants, sample storage, and how our plans fit with those of our hosts and host institutions.
- We must consider and assess potential risks to ourselves and those we work with.
- Pilot studies are vital to test methods, troubleshoot, and find out what is and isn't feasible.
- Pilot studies often reveal exciting new possibilities.
- Detailed timelines and budgeting are essential to ensure a project is feasible.
- Preregistration of research plans addresses problems of *post hoc* theorising, improves science, and has enormous potential benefits for our field.

With a feasible study design and a detailed research plan, we can now write funding proposals. In the next chapter, we look at how to do this.

17.8 FURTHER READING

Center for Open Science (COS). https://cos.io/. Includes useful frequently asked questions about preregistration.

Nosek BA, Ebersole CR, DeHaven AC, Mellor DT. 2018. The preregistration revolution. *Proceedings of the National Academy of Sciences of the United States of America* 115: 2600–2606. https://doi.org/10.1073/pnas.1708274114. A clear presentation of the differences between predictions and postdictions (*post hoc* theorising), the psychological biases which lead to mistake postdictions for predictions, and how preregistration avoids these problems. Also addresses common concerns with preregistration.

O'Donnell J. 2011. *How to make a simple Gantt chart.* https://theresearchwhisperer.wordpress.com/2011/09/13/gantt-chart/ [Accessed 9 January 2019]. Advice on project planning and on how to prepare a project timeline.

O'Donnell J. 2014. *How to make a simple research budget.* https://theresearchwhisperer.wordpress.com/2014/10/07/simple-research-budget/ [Accessed 9 January 2019]. Advice on how to create a budget for a simple project.

18

Writing a Research Proposal

Whatever you propose to study, you are likely to need funding for equipment, supplies, transport, and other expenses. You may also need to cover tuition fees and living expenses. When you apply for funding, you enter a competition. The strength of competition for funding means that your application must be excellent to have a chance of being successful.

In this chapter, I explain how you can improve your chances of obtaining funding by applying to appropriate organisations, tailoring your proposal carefully, following guidelines, and seeking feedback while preparing your application. I then provide general points on writing and details of each section of a proposal. I review the factors affecting success and how to deal with the outcome of the funding decision. I focus on relatively small grants appropriate for PhD students and postdoctoral scholars. Much of the advice also applies to PhD project proposals.

18.1 IDENTIFYING SOURCES OF FUNDING

Your first task is to identify appropriate funding sources. The funding available to you depends on your nationality, career stage, employment situation, and research interests. If you are affiliated with a university or institution, they may have internal funds you can apply for. Other sources include national research agencies, national and international learned societies, and non-governmental organisations.

Some universities have a research office that can help you to identify sources of funding. Colleagues, research funding databases, and Internet searches are other good sources of information. Check the acknowledgements of articles and conference presentations and researchers' websites to find out who funded the research.

18.2 PRIORITISING AND PLANNING

Once you have a list of funding sources, consider each one carefully. Check whether you are eligible to apply. Note the sorts of research activity they support. What size grants do they offer? Do they fund projects in your area? Look at current and recent grant holders and compare your CV with theirs. Do they fund people at your career stage? Note the deadlines, including the time of day and time zone. Note success rates, and details of the application process, including any documents, references, permissions, or letters of support you need to obtain. Check for restrictions that might make your project inappropriate (e.g. a fellowship may restrict the amount of fieldwork you can do). Get hold of successful applications if you can.

Once you have a list of potential funding sources, prioritise among them based on suitability and success rates and make a timeline for your applications. Good applications take time, and low success rates mean you will need to make multiple applications to support your work. Don't rely on just one application.

Share your application plans with your research partners, including potential or actual supervisors, mentors, hosts, and collaborators. Seek their advice. Include plenty of time for feedback in your funding plan. You need to allow your readers time to read and reflect on your proposal, and for you to address their comments. Don't expect your partners to be available at the last minute. Check for internal deadlines in institutions. These may be weeks ahead of external deadlines to allow time for institutional review and approval processes.

If your project involves research that seeks to address an applied research question, involve the people concerned and those whom you hope will use your findings (as stakeholders) in your project as partners. Providing target audiences with reports at the end of a project is not effective and stating that you will do so will not convince funders to fund your project.

18.3 UNDERSTANDING THE FUNDER'S PRIORITIES

No project, however great, will be funded without attention to the funder's priorities. Submitting a proposal that doesn't meet the funders' criteria wastes everyone's time and may damage your reputation. You must tailor your application to the funder's aims and follow their guidelines carefully. This means you can't use the same proposal for more than one funding agency, although you can use the same project.

To write a good proposal, you must understand the process from the point of view of the funder and the assessment panel. Scrutinise the funder's website to understand how the funding process works. Read any guidance they provide carefully. If they publish their evaluation criteria, read these criteria carefully. You need to address *all* of them *explicitly* in your proposal.

Understand the function of each section of the application and attend to all sections of the application, not only to the description of the research project itself. Funding decisions are often based on the other sections of an application, because all the projects themselves are excellent. Funders may publish the percentage weighting applied to each section of the proposal. Examine this carefully.

Find out who will assess your proposal. It may be assessed by academics or non-academics, specialist or non-specialist reviewers, or a combination of these groups. This matters for how you write the application, how much specialised knowledge you assume the assessors will have, what their priorities are, etc. Put yourself in your readers' position when you read your own proposal. Ask others to do the same when giving you feedback. You often need to convince critical, busy, assessors who are not specialists in your specific field of the value of your proposal.

18.4 SEEKING ADVICE

Writing a successful proposal takes time and care. Detailed, critical feedback is essential. It's much better to get this before you submit, when you can address it, rather than from the funder, when it is too late.

Peer review identifies flaws and weaknesses in your proposal and potential misunderstandings. Ask your readers to be critical and explain exactly which aspects of the application you want feedback on. Tell them that encouragement and supportive comments may be well intentioned but are far less useful than constructive criticism. Connect to other researchers who can provide feedback via your institution (if you have one), learned societies, conferences, and email requests (Box 13.2).

Seek specific feedback at all stages of proposal preparation, from the initial idea to a draft proposal. This includes your approach, project design, details of each section of the proposal, and overall clarity and logic. Seek advice from different types of reviewers and ask them to respond to specific questions about your proposal. Experts in your specific topic can advise on missing literature and the specifics of study design and methods. People from similar fields can advise on the overall approach. There may be general criticisms of your field that you need to address explicitly so that reviewers do not dismiss your application based on a differences in approach. People with experience of the funding competition can provide feedback specific to the funder's criteria. Non-specialists can help with clarity and logic and check that the language you use is comprehensible to non-experts. Ask readers to summarise your project. If they find this difficult, or if the summary is not what you expect, consider why and revise your application.

18.5 WRITING THE PROPOSAL

Assessors are likely to have many applications to read, so reduce to a minimum the work they need to do to understand your application. Your proposal must be easy to read, easy to understand, and convincing. Be consistent and logical and use a simple structure. Use clear, simple language (Chapter 6) and make every word count.

Structure your proposal so that overstretched readers can find and understand all the information they need to evaluate it in one quick reading. The assessors will not have time to search for information, so if you don't make your point very clearly, they will not get it. Make your key points early; don't make the readers wait to the end. They may not read that far. Test this by asking reviewers to read your application quickly and give you feedback on whether they could recall the information needed to evaluate your proposal based on the assessment criteria.

Address all the funder's criteria clearly and provide clear evidence to support a high score on all of them. It is not enough to state that you meet a criterion; show how you meet it. Check that you have done this against a checklist of the criteria.

Avoid excessive claims but don't undersell your project. It's okay to tell the reader that your project is exciting, cutting edge, state-of-the-art, unique, innovative, and world class, but only if you support all your claims with evidence for *how* your project is exciting, cutting edge, etc. Reviewers will disregard unsupported assertions. Feedback is useful here, too.

Be consistent with terms, limit your use of technical terms, and define any you do use clearly the first time you use them. Write in full, short sentences with correct spelling, grammar, and punctuation (Chapter 6). Write short paragraphs with clear topic sentences that summarise each paragraph, supported by detail in subsequent sentences. Ensure that the flow of ideas is logical and consistent. Provide context for each point you make; don't make readers wonder why you mention something. Explain the conclusions you draw from the material you include; don't expect readers to do this for you. Ask your readers to check this and mark any sections that are unclear.

Use informative headings that describe the content of a section. Uninformative headings (e.g. Experiment 1; Study 1; Phase 1) make it more difficult for your readers to follow your reasoning.

Use repetition to ensure that readers who skims the application grasp the point you wish to make.

Use clear figures to help readers understand your project.

Pay scrupulous attention to instructions, including font size, line spacing, and margins. Assessors may throw out an application that ignores instructions. Sloppy presentation will have a negative influence on the assessor's opinion of your application.

Funders often request a specific format. If they do, stick to it carefully and be sure to fill in the sections and answer the questions they ask carefully and thoroughly. Information must be in the right place so that the reviewers will find it when they expect it. Never just copy and paste in the closest text you have from another application. That will not work.

If the funder does not have a specific format, use the following structure, with subheadings.

18.5.1 Title

The title is your chance to give a good first impression. Take time to think about a short title that defines your project clearly and accurately and is appropriate for the funding body. A clever or humorous title is okay, as long as it doesn't hide the scientific content, but can be risky if the joke falls flat.

Use the title to tell readers why the study is important beyond the case study you propose, but don't overgeneralise and suggest you'll accomplish more than is possible. Cut all unnecessary words (e.g. *A study of* or *An investigation into*) and jargon, and don't use too many modifiers before a noun. Look at the titles of funded projects for inspiration.

18.5.2 Summary

The summary is very important. Write it last. Your readers need to know what your project is about straight away. The summary must immediately engage the readers' interest and convince them of the importance and feasibility of your project. State the main research question, and then explain why we need to answer it (i.e. its importance). List three to five specific objectives that will resolve the question and explain how you will achieve these. End with the significance of the study.

18.5.3 General Aim: The Problem You Will Solve

Begin with your **aim**. Your aim is the overarching purpose of your study. It describes what you intend to achieve. It is based on your research question. An aim is usually *to* something, for example, *to understand*, *to investigate*, *to determine* or *to test*. It must be clear, concise, ambitious, and achievable. Ask your readers for feedback on this.

Next, outline your three to five **objectives** as a numbered list. These are the specific steps you will need to take to achieve your aim. Use strong positive statements. Objectives should be realistic, clear, precise, and brief. They focus on outcomes, not processes, and act as milestones that you will achieve during your study. Objectives must be clearly related to your aim, achievable, and measurable (Box 17.1). They should not repeat the aim.

Get your essential message across quickly and simply to capture your reader's attention. By the end of the first half page of your proposal, they should be convinced that your project is exciting. The rest of your application fleshes out that message, backs up your claims, and shows that your plans are feasible.

18.5.4 General Background: Why the Problem Is Important

The general background puts your project into theoretical context. This is broader than your particular case study. Review the relevant literature to explain the importance of the problem you have identified. Summarise the current state of knowledge precisely and succinctly. Focus on what we do not yet understand (the gap in our knowledge) and why we need to understand it (the significance of the problem). Don't write a general literature review on the topic. Explain why the project is timely. Outline the key hypotheses you will test. Include key references; omitting one of these can wreck your chances of funding.

By the end of this section, your readers should be convinced that the topic is exciting, that you have an excellent grasp of the literature on the topic, and that addressing the problem that you set out to address would lead to a major advance in our understanding.

18.5.5 Specific Background: Why Your Project Is the Solution to the Problem

Begin this section by stating that you can solve the problem that you introduced. Explain your case study and why it is appropriate to address the problem. Summarise the approach you will use and why you (or your team) are the right person (or people) to address the problem. In other words, explain the insights, theoretical expertise, and skills that you bring to the project.

By the end of this section, readers should be convinced that you can answer the question you set out to answer.

18.5.6 Objectives: Specific Steps You Will Take to Address the Aim

Repeat your objectives and provide more detail to justify and explain them. Link them clearly to the aim. Objectives often include the specific hypotheses that you will test.

Your readers should be convinced that the objectives are achievable and will allow you to resolve the problem.

18.5.7 Project Plan, Methods, and Timeline: A Recipe for Success

Once you have convinced your readers that you have identified an exciting problem and that you can address it, you need to demonstrate that your project will succeed.

Provide a detailed project plan that is clearly related to your objectives. Include your hypotheses and predictions, and detail how you will address each one, including the study design. Justify all your methods thoroughly and link them clearly to your objectives. Provide evidence that the methods are appropriate and feasible. Be clear and logical. Include your intended sample size, show that it is achievable, and explain exactly how you will analyse your data to test your predictions, including the variables you will need in your analysis, how you will derive them, and the statistical tests you will use. Stating which statistical software you will use is not sufficient. Explain how you will interpret the results to determine whether your hypotheses are supported. Include any preliminary data you have that show your approach will work.

Include a clear, realistic, and detailed timeline to demonstrate that you can manage your project to completion (Section 17.4). Anticipate and address possible criticisms. Identify risks and provide contingency plans to address them. Show that your project will produce interesting results regardless of whether your hypotheses are supported. Include logistics, permissions, and preparatory work.

If your proposal involves career development (e.g. for a PhD or postdoctoral scholar), explain any training you will undergo to gain the skills you need.

This section should convince expert assessors that you can achieve your aims in the time available.

18.5.8 Ethics

Address the ethical implications of your project, including any permissions you have already obtained, or your plan for obtaining those permissions (Chapter 2). All proposals have ethical implications, so don't simply state that no such considerations apply.

18.5.9 Significance: What the Project Will Contribute

Return to the central question of your proposal and explain the contribution your project will make to big questions, and who will benefit. In other words, explain why people should care about the project. Make explicit reference to the goals of the funding body.

Someone who knows relatively little about your field should be able to understand this section and be convinced that the project is important.

18.5.10 Dissemination: What You Will Do with the Results

Explain in detail how you will communicate your findings. For example, you might aim to write and submit a specific number of manuscripts for publication in peer-reviewed journals, present at specific conferences, write a thesis, write a blog about your research for a general audience, write press releases for media coverage, present your findings to stakeholders, and hold a workshop to discuss them, or produce other **outputs**. In each case, identify your target audience and justify your choices. Be specific.

18.5.11 Track Record: Why You're the Right Person to Tackle the Problem

Funding applications may include a narrative description of your track record, or simply a CV. Both function to convince the assessors that you possess the skills and ability to conduct the project, so tailor this section to the application. Provide evidence of your achievements and highlight specific aspects of your experience that make you able to conduct the project.

If you don't have experience in a key area, then describe other aspects of your experience that suggest that you will be able to achieve your objectives, how you will acquire the necessary skills, or collaborate with someone who has those skills. For example, if your project involves collecting behavioural data on wild animals, and you don't have experience in this, you could combine experience with captive animals, or with video analysis of behaviour, with other evidence that you will adapt well to fieldwork to show that you are well prepared for your project.

Having a passion for primates and primatology is important personal motivation, but inappropriate as a justification for requesting funding. Instead, focus on your skills and experience.

CVs usually include:

- University qualifications
- A brief employment history (posts held, dates)
- Previous funding, including the agency, title, date, value, and whether you were principal investigator
- (Selected) publications. Publications are important evidence of peer-reviewed outcomes, and demonstrate your ability to complete a project
- Other dissemination (conference presentations, technical reports, public engagement)
- Prizes, awards, invitations, and other indicators of esteem
- Project and budget management experience
- Supervision of research students

- Relevant training in specialist research skills
- Other relevant experience and specialist skills, including non-academic experience, and an explanation of why it is relevant

18.5.12 Budget and Budget Justification: The Funding You Need

You usually need a comprehensive, detailed budget for a funding proposal. This may not be necessary for a studentship or fellowship, but if these have no or limited research costs you'll still need to explain how much your research will cost and how you will fund it. For larger grants, you will need help with your budget and must include the time this takes in your plan.

Your budget should be realistic, adequate, and accurate. In other words, it should match the project. Reviewers will check this. Do not skimp; reviewers will spot the mismatch between your aims and your budget. Check that none of the costs appears excessive and that the costs you include are eligible.

Follow the guidelines and justify each item carefully, showing exactly why the costs are necessary, and that the project is cost-effective.

Include other sources of funding, both secured and pending. Explain whether you will be able to conduct the project if your other applications are unsuccessful.

18.5.13 Bibliography

Some application forms include a bibliography section; others do not. If there is no dedicated space, then place the bibliography at the end of the relevant section(s). If the word count is limited, use a superscript number for citations (like this[1]). If you have space, replace this with an (author date) format (like this: Darwin 1859) so that readers do not have to refer to the bibliography to see authors and data of publication. Use a consistent bibliographical style. If space is limited, omit line breaks and article titles.

18.5.14 Other Documents

Arrange all the documents, references, or letters of support (Box 18.1) that you need to obtain well in advance and provide your project partners with all the information they need. Never include collaborators without their agreement.

Box 18.1: Asking for a reference letter

We often need reference letters, or letters of recommendation, for applications for funding, scholarships and positions. In some cases, reference letters determine which applications succeed and which fail, so you need the letters to be as strong as possible.

The ideal referee understands you, the application, and the organisation you are applying to, and knows how to write a good reference letter. He or she should have relevant expertise and be positive about your application. Some organisations specify the position the referees should have and their relationship with you.

Request reference letters at least 3 weeks in advance of deadlines and take national holidays and particularly busy periods (e.g. university marking periods) into account. Your referees will have many other responsibilities and may not always be available. Even if they can fit in a last-minute request, the reference will not be as good as it would be if they have plenty of time to prepare. Don't rely on requests sent by the organisation you are applying to; these may go to your referee's junk folder. Send your referees a polite reminder a few days before the deadline.

Explain to your referees why you have chosen them. If you haven't been in contact with them for a while, provide them with an update on what you have been doing, including a summary of your current work, future plans, and why you are applying for the funding or position. Consider whether they are still the best people to recommend you, if you have gained experience since they mentored or advised you, or since you last worked together closely.

If you intend to ask the same referees for several letters, provide them with a list of deadlines and requirements so they can plan in advance.

The people you ask may decline to write a reference letter, if they feel that they cannot write a strong letter to support your application, don't know you well enough, or don't have time.

If your chosen referees agrees to write a letter, make their task as easy as possible. They need to know what your application is for, why you are applying, and why you asked them to recommend you. Provide them with all the information they need, in a single package (usually in an email), so that none of it goes astray. Provide them with a description of the funding opportunity, programme, or

Box 18.1: (cont.)

position you are applying for, any instructions for referees, your application, and any accompanying documents. If there are forms for the referees to fill in, provide them, and complete as much information as you can yourself. (Do not expect them to fill in your details, for example).

Provide your referee with your CV and highlight any achievements, experience, or training that are particularly relevant to your application. Provide specific examples and illustrations of strengths you wish to highlight. This should include any criteria that referees are asked to address. This can feel like boasting, and make you feel uncomfortable, but it is necessary. Don't assume your referees can remember every detail about you and your accomplishments. Don't rely on your referees to find information, particularly if it is from some time ago.

If you are concerned about any weaknesses in your application, discuss these with your referees. They may be able to help.

Reference letters are often subject to unconscious (implicit) bias based on gender, ethnicity, or other aspects of identity (Chapter 4). If you feel comfortable doing so, point your referees to good practice guidance on writing letters.

Finally, remember to say thank you to your referees for reference letters, and let them know the outcome of your application.

FURTHER READING

Dutt K, Pfaff DL, Bernstein AF, Dillard JS, Block CJ. 2016. Gender differences in recommendation letters for postdoctoral fellowships in geoscience. *Nature Geoscience* 9: 805–808. http://dx.doi.org/10.1038/ngeo2819. Analyses 1224 recommendation letters for postdoctoral fellowships in the geosciences, submitted by recommenders from 54 countries, for 2007–2012. The findings suggest that women are significantly less likely to receive excellent recommendation letters than men at this critical career stage.

Ross DA, Boatright D, Nunez-Smith M, Jordan A, Chekroud A, Moore EZ. 2017. Differences in words used to describe racial and gender groups in Medical Student Performance Evaluations. *PLOS ONE* 12: e0181659. https://doi.org/10.1371/journal.pone.0181659. Investigates standardised evaluations made at the transition from medical school to residency. Reveals systematic differences in how candidates are described based on racial/ethnic and gender group membership.

Schmader T, Whitehead J, Wysocki VH. 2007. A linguistic comparison of letters of recommendation for male and female chemistry and

Box 18.1: (cont.)

biochemistry job applicants. *Sex Roles* 57: 509–514. https://doi.org/10
.1007/s11199–007-9291-4. Uses text analysis software to analyse letters
of recommendation, finding that letters included significantly more
standout adjectives to describe male candidates than female candidates.
Inspired Thomas Forth to create the Gender Bias Calculator (www
.tomforth.co.uk/genderbias/).

Trix F, Psenka C. 2003. Exploring the color of glass: Letters of recommendation for female and male medical faculty. *Discourse & Society* 14:
191–220. https://doi.org/10.1177/0957926503014002277. Analyses all
the letters of recommendation for successful applicants for faculty
positions over 3 years in the mid-1990s at a large US medical school.
Shows systematic differences between letters for women and men, to
women's disadvantage.

18.6 FACTORS AFFECTING SUCCESS

Funding agencies cannot fund all the high-quality proposals they
receive. This means that high-quality applications are often rejected,
and assessors rely on minor points to rank proposals. Don't give them a
reason to reject yours.

An excellent proposal:

1. Identifies a clear, compelling problem that is relevant to the
    funding body and justifies why it is important.
2. Proposes a well-designed project that will answer the question.
3. Shows that the project is feasible, with a clear and realistic
    timeline, and an appropriate budget.
4. Demonstrates that the applicant has appropriate track record,
    with the skills and experience required to conduct the project.
    (If you don't have experience in a key area, then describe other
    aspects of your experience that suggest that you will be able to
    achieve your objectives.)
5. Provides clear evidence to support high scores on all the funder's
    criteria.
6. Is well presented.

18.7 HANDLING THE OUTCOME

The time from application to outcome varies from weeks to months.

Most applications are unsuccessful. Some of the factors affecting
success are out of your hands, including the strength of the competition

and bad luck. The strength of the competition can mean that an application passes the threshold for funding, but funds are limited and other proposals score higher, so your application is unsuccessful. Bad luck may mean that a funder must choose between many excellent proposals and you are unlucky.

Plan for rejection, because it is almost inevitable. Don't pin your hopes on a single application. Instead, apply for several times the amount of money you will need and plan time for resubmission if the funding agency allows this. Remember that rejection is not a comment on you personally, but on the application, and that the best scientists get rejected.

Read any feedback you receive carefully, stay positive, and learn from it. You may be able to resubmit your application to the same funder, or to an alternative funding source, but you also need to know when to ditch a project if the reviewers highlight problems you cannot resolve. You may disagree with the reviewers' comments, but you still need to learn from them. If the reviewers didn't see the importance of your question, or misunderstood your proposal, then you need to revise it. If you resubmit, address all the reviewers' concerns. Don't resubmit the same proposal.

If you're successful, celebrate! Thank those who helped you. Use the funds solely for the purposes specified in the award and follow all guidelines. You may need to keep detailed accounts and receipts for all expenditure. Submit all progress and final reports required on time and comply with all guidelines.

If you don't get funding for a project, you may need to revise it, or abandon it in favour of a cheaper option.

18.8 CHAPTER SUMMARY

In this chapter, we've seen that:

- Funding is often highly competitive and even excellent applications may not be successful. Good is not enough.
- Writing winning funding proposals takes good planning and a lot of time. We can't do it at the last minute.
- Attention to the funder's priorities is crucial; we can't copy and paste from one application to another.
- We can improve our chances of obtaining funding by learning to read an application like a reviewer.

- Critical feedback is essential to identify flaws in a proposal; we must plan time to address them.
- Grant assessors often work under pressure, so we must make their task as easy as possible.
- It is not enough to state that a project meets a criterion; we must explain how.
- We shouldn't pin all our hopes on a single application.

Box 18.2 addresses common problems with funding proposals and how to resolve them.

Box 18.2: Common problems with funding proposals and how to resolve them

Common problems with funding proposals include the following points:

The Proposal Is Not Compelling
If your application fails to explain the research problem to the appropriate audience, reviewers will be left asking *so what?* or *who cares?* Don't rely on your reviewers to figure out why your study should interest them. This often happens because the focus is too specific, and you don't explain the general implications for those not interested in your particular case study. Check for this by seeking advice from someone who is not familiar with your research topic.

Linked to this is a proposal that assumes knowledge the readers may not have. Find out who will review your application. You don't need to explain primates to primatologists, but will need to justify the choice of primates as a case study when applying for grants that are not specific to primates. Reading popular articles about your research topic is a useful way to understand the level of understanding you can expect your audience to have.

Ineligible Proposal or Poor Fit to the Funder
The funder has criteria and priorities. Your proposal will not be funded if it does not fit these, however good it is. You must highlight how your proposal fits the criteria. You cannot rely on reviewers to figure this out for themselves. This means that copying and pasting material from a different proposal will not work.

Box 18.2: (cont.)

To resolve this, read the guidelines, the funders' aims and priorities, and their evaluation criteria carefully. Address all the criteria explicitly in your proposal. This may involve changes to your original plans. Give the criteria to your reviewers when you request peer review and ask them to check whether you address them.

Flawed Proposal

Flaws include a lack of critical review of the literature, an unsound hypothesis, overly ambitious plans, a flaw in the study design, inappropriate methods, inadequate sample size, inadequate description of methods, problems with the budget, or a lack of evidence that you will be able to conduct the study. It's much better to have readers pick up on these flaws before you submit than to wait for reviewers to find them when it's too late to fix them.

Inconsistent or Vague Proposal

Inconsistent proposals have no clear link between sections. For example, the introduction might set up a problem, but the study design doesn't address that problem. Similar problems include hypotheses that come as a surprise to the readers rather than being clearly linked to the introduction, methods that are unrelated to the aims, and budgets that are not linked to the rest of the proposal.

Vague proposals include those where the introduction reviews a general topic but does not lead clearly to the aims of the project, the methods are too imprecise to assess whether they will address the aims successfully, the budget is not adequately justified, or the CV does not clearly demonstrate that you can conduct the project. Fix these issues by focussing the introduction on your aim, providing clear methods including projected sample sizes and details of exactly how you will test your predictions, a clear justification for your budget, and a CV tailored to the application.

Dense or Incomprehensible Text

Poor writing can kill a proposal. Reviewers may not have the time and inclination to figure out what you mean and will favour a well-written, competing proposal. Follow the advice in Chapter 6 and seek help if you know you have difficulty writing clearly.

Box 18.2: (cont.)

Material in the Wrong Place
If you include material in the wrong part of an application, reviewers may miss it. They may score each section of the application, meaning that information that appears in the wrong place doesn't count.

The Application Repeats
Repeats of key messages are useful, to ensure that a reviewer gets the point. However, repeating more than short phrases bores the reader and wastes your word count. Ensure that the material in each section is relevant to that section.

Failure to Follow Instructions
If you don't follow the instructions for the proposal, reviewers may doubt that you will conduct good science.

Project Is Not Feasible
Projects may not be feasible because they are not achievable within the proposed time period, or with the money available. To resolve this, ensure that your proposal includes appropriate time for all phases of your project, including set-up, data collection, analysis, and reporting. Include a timeline that demonstrates this clearly to your reader (Section 17.4). Also ensure that the budget is appropriate for the work, justify each budget line clearly (Section 17.5), and check that it complies with the funder's instructions.

With a detailed proposal, and (hopefully) funding, we can collect data to address our research question. The next chapter looks at data collection.

18.9 FURTHER READING

Aldridge J, Derrington AM. 2012. *The Research Funding Toolkit: How to Plan and Write Successful Grant Applications.* London: Sage Publications. A handbook for how to write grant proposals. UK-focussed but includes international examples. Very useful for those applying for larger grants and includes advice on large-scale collaborative projects. Accompanied by a website and blog at: www.researchfundingtoolkit.org.

Clutton-Brock TH. 2000. *Survival Strategies for Scientists.* www.zoo.cam.ac.uk/research-groups/larg/tim-clutton-brock-survival-strategies-for-scientists/view.

[Accessed 9 January 2019]. Tips from Tim Clutton-Brock, of the Large Animal Research Group at the Department of Zoology, University of Cambridge, UK, on grant and fellowship applications.

Hailman JP, Strier KB. 2006. *Planning, Proposing, and Presenting Science Effectively*. Cambridge: Cambridge University Press. Chapter 2 covers how to write a research proposal.

Karban R, Huntzonger M, Pearse IS. 2014. *How to Do Ecology: A Concise Handbook*. 2nd edn. Princeton, NJ: Princeton University. Chapter 8 includes advice on writing grant proposals.

Also look out for blogs and websites with advice on writing funding applications.

19

Collecting Data

Data collection can be fun, but it can also be difficult. Things don't always go to plan. Ask other researchers about projects that didn't work; we all have plenty of examples.

In this chapter, I cover the importance of monitoring the progress of your project, being flexible and open to opportunities, being prepared for the unforeseen, collecting data rigorously and systematically, keeping data and samples safe, and being considerate of other people.

19.1 KEEP TRACK OF YOUR PROGRESS AND YOUR SPENDING

Always keep an eye on your priorities. It is very easy to be distracted from the data you need to collect to test your predictions. If you decide, or need, to change your plan, assess the advantages and disadvantages of the changes in the context of your study aims and objectives.

Monitor your progress weekly. Review and update your timeline as your project progresses. If things go well, you may be able to add additional data collection to your project. If they don't, you need to identify problems and revise your plans. Tracking what you've done is also very useful when you're feeling overwhelmed, because it provides a record of what you've already accomplished (Box 1.3).

Track your spending and compare your records with your budget. If you're running low on funds, prioritise the most important tasks, but remember to use funding only for the purposes specified in the award. You may need to collect and submit receipts. If your plans change, inform the funding body and ask their advice about reallocation of funds.

19.2 BE FLEXIBLE AND OPEN TO OPPORTUNITIES

Although planning is essential in research, we must also be flexible and open to new avenues of research depending on what we encounter. Few, if any, projects follow the original plan. Instead, they change direction, sometimes multiple times, and sometimes drastically. Serendipity plays a major role in research, so we need to balance a clear focus on our question with the ability to change plans when a more interesting possibility suddenly presents itself. In other words, we must focus on our goals, but we must also expect the unexpected, improvise, adapt, and take advantage of opportunities.

19.3 BE PREPARED FOR THE UNFORESEEN

Even if your pilot study went well, there will still be some things you haven't thought of, and many things can go wrong with a plan. For example, there may be no (or very few) infants born in a given year when you plan to study infants, animals may not call so you can't record them for a project that relies on acoustic samples, animals may not enter traps so you can't catch them, animals may remove tracking collars, or collars may not work, and animals may fail to interact with experimental apparatus.

If you work with captive animals, you must work within the constraints of the facility, including their daily husbandry schedules and opening/closing hours (Section 17.1). The needs of your study animals or your host facility may interfere with data collection. For example, the group of animals you're working may be split, animals may be sent to another facility, or animals may be placed on contraception (inconvenient for a study of reproduction). Maintain good and clear communication with your host to help ensure that your research needs are at the forefront of peoples' minds when they make decisions that may affect your research.

Remember your contingency plans (Section 17.4). If everything goes wrong, and you can't collect any of the data you need, make another plan based on what is possible. It's much easier to develop a research plan when you've done it before. This is where your broad background reading comes in (Section 10.4). Remember to go through all the decisions you need to make when designing a study and justify each one (Box 17.2).

19.4 COLLECT DATA RIGOROUSLY AND SYSTEMATICALLY

Collect data carefully and methodically. Use blind protocols if possible (Section 3.3) and test the reliability of your methods (Section 14.3). Practice your methods to improve intra-observer reliability and reduce variation in measures of the same material (**observer error**). Observer error can be due to inexperience, drift in methods over time and mental lapses due to fatigue. Don't change your methods during data collection; otherwise your data will not be comparable over time.

Train any research assistants carefully. People may differ in how they record data, so establish and check inter-observer reliability (Section 3.3) regularly by making the same measurements independently, comparing the data and measuring how well they agree. How you do this depends on the data you're collecting, but you need to know the level of agreement, not the statistical significance. For example, you might assess the number of agreements, disagreements, and omissions in categorical measures using Cohen's Kappa statistic, or compare continuous measures made by two people using the interclass correlation coefficient (for more than two people use intra-class correlations). If the data are similar, you can proceed. If they are not, you need to check your methods and retrain the observers. If you still don't obtain data that agree, you may need to rethink your measures.

Be rigorous but remember that no study is perfect. If you are a perfectionist (many scientists are), beware of being over-rigorous. No study is perfect. Concentrate on what you can achieve. Don't get hung up on data you can't collect.

19.5 DON'T PEEK AT YOUR DATA

Set a predetermined stopping point for data collection, based on the sample size you need. Don't test your data for statistical significance as you go along, to avoid problems associated with data-peeking (Section 3.5). If, for example, you collect subsets of data and test your predictions periodically (e.g. after each field season, set of experiments, or data collection trip) using null hypothesis statistical testing, and stop collecting data when you detect the effect you were hoping to detect, you increase the risk of reporting a false positive due to random variation in your data. Either leave your data analysis until you have finished data collection or use statistical analyses that account for data-peeking.

19.6 KEEP YOUR DATA AND SAMPLES SAFE

Keep clear, legible, and understandable records of data collection. You will forget what any codes and abbreviations you use mean, so keep a record of them.

Organise your data and files, and back them up carefully. Data on paper can be lost, damaged, stolen, eaten by animals (from ants to elephants), or end up at the bottom of a river. You can photograph hard copies of data and save the images with searchable file names but entering the data into a spreadsheet straight away, while you are familiar with the data, will save time in the end. Back-up electronic files and keep multiple copies in different locations, ideally in the cloud. This applies at home as well as in the field. A cup of coffee or a domestic cat can put your computer beyond repair.

Label samples indelibly and catalogue them carefully with all the information you need. Check stored samples regularly for lost labels, temperature of cold storage, fungal growth, and other problems. If possible, store duplicates of your samples elsewhere.

19.7 WORKING WITH OTHER PEOPLE

Working relationships require respect, sensitivity, and openness to other people's opinions (Section 2.8). We must respect the priorities of the people we work with, who might include technical staff, veterinarians, animal care staff, local guides, park officials, other researchers, students, and research assistants. Be courteous and helpful and maintain positive relationships. Communication is very important, and dialogue can help to find solutions to problems. Concealing frustration and responding well to negative feedback are also important. Never assume that you are more knowledgeable than the person you are talking to. We each bring our own competencies and experience to a working relationship.

Educate yourself about the definition, prevention, and reporting of harassment and abuse (Section 4.11). Consider safe, responsible, and effective ways to react or intervene if you witness or are informed about such behaviour.

If you employ people, or work with volunteers, train and mentor them appropriately and ensure your expectations are reasonable. Be aware of the power relationship you have with them.

You may come across the public during data collection, for example, if you work at a zoo, or if tourists or other people visit your field site. This is an excellent opportunity to explain your research and to be an ambassador for primates and primatology.

If you work at a zoo, make it clear that you are a researcher. For example, you might wear a T-shirt or badge identifying you as a researcher. This will help members of the public to understand that they shouldn't copy you, for example, if you use an area that the public are not allowed to use, or if you interact with the animals. As we saw in Section 17.1, you might need a sign to say that you are collecting data.

19.8 CHAPTER SUMMARY

In this chapter, we've seen that:

- Data collection is often fun but can be difficult and rarely goes to plan.
- We must focus on our goals and monitor our progress towards them, but we must also be flexible, and take opportunities that present themselves.
- We must collect data rigorously and blind to the question if possible.
- We must have a predetermined stopping point for data collection and shouldn't peek at our data as we collect it.
- We must keep data and samples safe.
- We must maintain good communication with collaborators and hosts.

Each research location brings its own challenges for data collection. Fieldwork, however, brings the greatest challenges, both physically and psychologically. In the next chapter we look at these challenges.

19.9 FURTHER READING

Caro TM, Roper R, Young M, Dank GR. 1979. Inter-observer reliability. *Behaviour* 69: 303–315. http://dx.doi.org/10.1163/156853979X00520. How and why we measure inter-observer reliability. Focusses on measuring behaviour.

Holman L, Head ML, Lanfear R, Jennions MD. 2015. Evidence of experimental bias in the life sciences: Why we need blind data recording. *PLOS Biology* 13: e1002190. https://doi.org/10.1371/journal.pbio.1002190. Shows that blind data collection is rare in the life sciences, provides evidence that this leads to bias, and proposes methods to address this.

20

Conducting Fieldwork

Fieldwork can be exciting, and even addictive, but it can also be daunting and dangerous. Field conditions range from a tent to established research stations. You may be close to home, or on the other side of the world. National researchers may be as foreign to a local area as non-national researchers are. You may be in a familiar environment or in a very unfamiliar one. Fieldwork often involves sharing living space with other people, and with wildlife.

In this chapter, I begin with what it takes to be a fieldworker, then cover permissions and logistics, field kit, personal safety, the social context, LGBTQIA+ concerns, natural hazards, physical health, mental health, and returning home.

20.1 WHAT DOES FIELDWORK TAKE?

Fieldwork doesn't suit everyone, and you don't have to do fieldwork to be a primatologist. Be honest with yourself, get some experience if you can, and decide whether you want to do fieldwork.

Fieldwork is not travel; it is living away from home. It changes you, often radically. It can challenge you to reflect on your own culture, assumptions, and expectations, and to learn new ways of seeing the world. It is an adventure that requires perseverance and passion. Fieldworkers need to be enthusiastic, independent, resilient, humble, patient, flexible, and tolerant of ambiguity when working in an unfamiliar setting.

Fieldwork requires both a very strong work ethic and adaptability – character traits that don't always go together. Similarly, fieldwork often requires the ability to live and work very closely with people but also to spend long periods alone. It requires a good sense of humour, initiative, strong motivation, and self-reliance.

You can conduct fieldwork solo, with peers, or with a partner (either as a collaborator or as support). Many primatologists take their children to the field, others prefer not to, while still others conduct short field trips to avoid being away for too long at any one time. If you're considering taking children with you, consult researchers familiar with your study site and researchers who have done so, preferably in the same location. Read this chapter with your children in mind, including childcare, medical care, insurance, and an evacuation plan. If you conduct fieldwork while pregnant, plan for medical care you might need in the field.

20.2 PERMISSIONS

Fieldwork requires permits and permissions at many levels (Section 2.2). Find out the deadlines for applications, how long it takes to gain permissions, and plan time to address any issues that arise with your application.

International fieldwork requires visas and permission to conduct research, capture animals where relevant, import and export equipment, collect samples, export and import samples. These permits and fees are one way for host countries to benefit from the research that you do there as their guest. Some countries will allow you to enter on a tourist visa and then negotiate research permission; others will not and ignoring this costs time and money.

Fieldwork in your home country requires many of the same permissions as international fieldwork, including research permission, permission to capture animals, to sample, and to travel with samples.

You may need permission from the local community where you conduct your research; you can't rely on national permits for this. Seek advice from local institutions and experienced researchers about both formal and informal permissions and allow plenty of time to obtain these.

Obey the law. This sounds obvious, but some fieldworkers seem to think it's optional.

20.3 LOGISTICS

Fieldwork logistics include where to stay, how to get around, how to communicate with the outside world, how to access money and supplies, equipment and how to repair it, how to keep data safe, and how to store and transport samples. Seek advice from local contacts and colleagues with experience in the area. A pilot study is extremely useful to understand these issues (Section 17.3). If this is impossible, then include time in your timeline to familiarise yourself with your field site. Remember that you need a plan A, a plan B, and a plan C (Section 17.4) and be flexible.

You might work at an existing field site or set up your own. Maintaining a field camp, accessing safe water, washing clothes, and preparing meals can all take time, depending on your field set-up. Food may be very different to what you're used to and limited in variety. Supplies can be irregular. Pack treats for yourself and for your colleagues. Spices and herbs are lightweight and can enliven dull meals.

Field sites can be remote, and inaccessible during some seasons, or easy to access. Getting to your field site can take days of arduous travel or be as simple as a train journey. You may need a vehicle, or you might use public transport. If you're travelling to a remote site, take enough supplies, including food and medical supplies, to allow for delays. You may need to learn basic vehicle or boat maintenance.

Communications are improving, and even remote field sites have Internet connections or mobile/cell phone coverage, but this is not always the case. Satellite phones work where terrestrial cellular service is unavailable, but handsets are expensive, and calls may be, too. You may be able to use a radio to communicate with a local base station.

Investigate how to access money. If the local currency has a high exchange rate, you will need a way of stacking the huge pile of bills discreetly. Carrying your own receipt books is very useful if you need evidence of expenditure.

Find out what supplies you can buy close to your field site and what you need to bring with you. Arrange to buy supplies locally if possible. Airlines have strict regulations on what you can transport and on packaging.

Your equipment is almost certainly not designed for your field site. It will be sensitive to temperature, humidity, sunlight, dust, insect attack, and travel on bumpy roads. Consider what you would do if it failed. Find out whether you can get the power you need. Batteries are also sensitive to field conditions and may not be available locally. Make sure you have all the cables and adaptors you need. You may need surge protectors to protect equipment from power surges. Test your equipment and know exactly how it works before you go to the field. Read the manuals and take them with you. Take back-ups, spare parts, and a basic tool kit for repairs. Indicating silica gel beads are useful for drying things out. Heat them to reuse them when they change colour.

Your study animals, and other animals, might wreck test apparatus, cages, and so on, so be prepared to fix them.

Duplicate samples if you can and leave one set behind when you transport or ship them. This is good practice if you have local collaborators and provides you with a back-up. If you freeze your samples, check regularly that the freezer is functioning correctly.

Store food carefully. It attracts animals.

20.4 FIELD KIT

Appropriate field kit, including footwear, varies with the field site, so get advice from colleagues experienced with your study area.

You'll usually need quick-drying, neutral-coloured (i.e. beige and khaki) clothes. Check with experienced colleagues before choosing military-style clothing because wearing such clothing can lead people to assume you are in the military, causing misunderstanding at best and physical danger at worst. You'll need long sleeves and long trousers to avoid insect bites. You might need boots or gaiters to protect you from snakes, or leech socks. Good insoles improve the fit of waterproof rubber boots. Don't wear dark blue or black in areas with tsetse flies. Pockets are essential, especially those with zips (women's trousers often dispense with pockets or reduce them to useless dimensions). You may lose weight in the field, so take a belt. Take smarter clothes for meetings with officials, social events, and special occasions.

Keeping your kit dry can be a major undertaking in the field. Cotton (including underwear) won't dry in humid conditions and goes mouldy. Take good care of your feet.

Experiment with ways to carry your field kit to find a comfortable system. Waist packs are invaluable. Some people find fishing vests useful. Camera and binocular slings and harnesses will save you neckache by distributing the weight of your equipment but these can interfere with one another. You may need to keep your camera and other equipment dry in torrential rain. If your equipment does get soaked, resist the urge to turn it on to see if it still works. Instead, put it in the sun and let all the water evaporate from inside before you turn it on.

Take a simple waterproof watch with long-lasting batteries. Durable and weatherproof field notebooks are invaluable, as are waterproof, airtight containers, and drybags. Other useful kit includes an e-reader (so you never run out of reading material), ziplock bags, a multi-tool, a good torch (flashlight), portable solar-powered charging equipment, and lighters (you can't take lighters on aeroplanes). If you're studying nocturnal primates, you'll need a strong torch with a red filter to avoid damaging their eyes.

20.5 PERSONAL SAFETY

Fieldwork can be postponed or disrupted by civil unrest, rebel groups, and crime. Anywhere can be dangerous but you may have reduced access to help when in the field and emergency systems may be unreliable.

Assess the risks prior to travel. Universities often subscribe to useful risk websites. Also check government websites and ask local contacts. Situations can change over time; if you haven't visited a location for a while, get up-to-date advice. If violence does erupt, it is often very localised, and it may be safer to stay put than to travel.

You may be particularly at risk initially, when you are unfamiliar with a place, but, equally, you can become too relaxed when you feel at home. Be aware of local political and cultural events. Establishing close relationships with people at or close to your study site will help but be aware that your concerns may differ. You may face the dreadful situation of leaving them behind if you are evacuated.

Distribute your itineraries to key contacts, establish check-in times, and ensure these are feasible to avoid false alarms. Obtain official permissions and carry them with you so that you can show them to the authorities if needed. Get appropriate insurance. Travel insurance may not cover all the eventualities you need to cover. Expedition or university insurance is often better. Have an emergency plan and share it with your institution and emergency contacts. Register with your embassy or consul.

Road accidents are common in low- and middle-income countries and help is not always at hand. Vehicles and boats present potential safety hazards.

Kidnapping and murder are rare, but they happen. Theft can happen anywhere, but differences in wealth may be larger in the field. Moreover, fieldworkers sometimes carry a lot of money and equipment. This may be an unavoidable risk, but consider how you are perceived and do your best not to establish routines that can attract thieves.

20.6 THE SOCIAL CONTEXT AND CULTURAL UNDERSTANDING

Fieldwork usually involves working and living closely with other people. You may share a field camp or station with people you wouldn't otherwise choose to spend time with. You may not have the privacy you are accustomed to. This can foster deep friendship but can also lead to serious interpersonal problems including alienation and misconduct, with very serious implications for the victims.

Discuss the safety of potential field sites with trusted colleagues. Request a code of conduct and sexual harassment policy from potential sites. Field sites run by institutions should adhere to that institution's code of conduct in the workplace and have a reporting structure for incidents. However, surveys suggest limited awareness of such mechanisms, and that few victims who report harassment or assault are satisfied with the outcome of the report.

Field camps or stations are often multicultural, with researchers, assistants, and staff from very different backgrounds, further complicating interpersonal relationships. Being open-minded, respectful, flexible, and willing to examine your own assumptions all help greatly, as does having a sense of humour.

Fieldwork also often involves living and working in a culture different to your own, providing a life-enriching experience as well as endless possibilities for misunderstandings. Learn as much as you can about your field site and the surroundings before you go, including the language, history, and behavioural and cultural norms. Ask about appropriate dress, norms of politeness and boundaries. For example, cultures vary in how long one spends on greetings and small talk, pointing may be rude, language may be blunt and candid, or people may be unwilling to say *no*. Cultures also differ in their expectations of sharing resources and gift-giving, so find out how things work at your field site. People may give you gifts and may expect gifts in return. Cultures also differ in whether drinking alcohol is socially acceptable. Find out and respect cultural norms concerning public displays of affection.

If something wouldn't be judged acceptable at home, you shouldn't do it at your field site.

You may disagree with local norms, but it is respectful to conform to them as much as possible. Be aware of the potential for misunderstandings. Similarly, you may disapprove of locally accepted practices, such as hunting, but remember that you are a guest at your field site.

Consider the assumptions people might make about you based on your appearance, how you speak, or where you come from. For example, if you are, or look like you might be, from a nation that colonised your host country, this may have implications for how people interact with you. Similarly, if you come from a different area of the same country, you may be subject to stereotype, and city-dwellers may be viewed with suspicion by rural people.

Consider the expectations people may have of you. For example, you may receive requests for assistance that you are not qualified to address, such as medical advice. People may assume you are very wealthy, and often you are in relative terms. Be prepared for people asking you for money and equipment.

You may encounter shocking poverty in the field and be moved to help. If you are, do so carefully. Consult with the local community about their needs. Be realistic about the skills and resources you can contribute. Tangible, sustainable development work takes time, skill, and, above all, good relationships and an understanding of the local context. There are no quick fixes. Don't assume you can fix problems and don't present yourself as a *saviour* to local people. Don't make promises you may not be able to keep.

Do not hand out sweets (candy) to children. If you want to give gifts to children around your field site, ask their parents and teachers what would be useful and give gifts to them, not directly to children. In other words, don't turn children into beggars.

Find out how best to employ people if you need to do so. Mismatched expectations can lead to resentment. Investigate the history of research and researchers at your field site; this sets the scene for your own work. Similarly, your own conduct will influence the experience of future researchers. You are an ambassador for primatology.

Communication is vital. People may be suspicious of your motives, particularly if you remove samples from the field. From an external point view, primatologists do very strange things and our behaviour is easily misconstrued. For example, we may show interest in species that are regarded as pests or appear to value such animals more than we do people. We often collect primate faeces and other samples, and our explanations for doing so may not satisfy our interlocutors. We may catch and handle animals. We may perceive our work as being for a higher goal of scientific knowledge, but our research also brings us obvious personal benefits (degrees, possibilities of employment or career progression). Understanding this and establishing good communication with influential people in the local community can help. Also consider the influence that your presence and activities have on the local social and political environment (**researcher influence**).

Think in advance about what you will do if you come across illegal activity, such as illegal hunting and logging. It can be very tempting to set yourself up as some sort of wildlife warden, but you are not a law enforcement officer.

20.7 LGBTQIA+ CONCERNS

Fieldwork can be difficult for LGBTQIA+ people due to prejudice among other researchers and the cultural context at field sites. These prejudices and their consequences may differ depending on your gender presentation: male and female presenting/perceived people face different expectations, and different consequences for challenging them. Moreover, some countries have laws or cultures that make fieldwork dangerous if LGBTQIA+ people are open, or outed. In these contexts, the ease with which you pass among colleagues and by local cultural standards may modulate risks. Of course, this doesn't mean that everyone you meet in the field will be anti-LGBTQIA+ and having a close ally in the field with whom you can be yourself is very important. Nevertheless, you may need to consider keeping important aspects of your identity and life secret. Being closeted is emotionally demanding and is also complicated by the risk of inadvertent outing by friends or colleagues.

Online identities are publicly available and have global reach. Friends and colleagues in the field will probably invite you to connect on social media. If you are out online, then you may need to think carefully before following or friending someone or consider using filters to mediate publicly available information.

If these issues affect you, investigate your particular field situation and consult with other scholars on how to mitigate any risks. Research both the laws and the customs of the countries you are considering carefully and decide whether to go there and how open to be if you do go. Understanding the context in which the laws were enacted is important: some countries still have the old anti-sodomy laws on the books, and some countries have recently passed new anti-gay laws. Think carefully about how much risk you are willing to take and how possible it is to reduce that risk.

You can't assume that your university, site manager, or advisor will be informed on this issue. Talk to LGBTQIA+ researchers who have been to your site or to colleagues from that country. The American

Society of Physical Anthropologists' gAyAPA (queerbioanth.org) comprises LGBTQIA+ researchers who have conducted fieldwork all over the world and aims to support LGBTQIA+ scholars and allies in Biological Anthropology.

If these issues don't affect you directly, they will affect your colleagues, students, or assistants. Be an ally.

20.8 NATURAL HAZARDS

Days in the field can be arduous. You may have to splash through streams, cross rivers, or wade through swamps. You may be constantly wet, or covered in dust or mud, depending on the season. You will walk into spiderwebs and creepers with sharp thorns at face height. TV presenters may appear to swim through forests, touching all the plants with awe, but this is a bad idea – many plants are protected by ant colonies, thorns, or tiny hairs that get into your skin and itch. Some harbour wasp nests.

You will get lost, either following your study animals, or looking for them. Plan for this. Know the trail system, if there is one, and carry a map, if you have one. Carry a compass and know how to use it to get back to camp. Don't rely on a GPS, because the batteries will fail. Carry a whistle. In mature forest, you can drum on tree buttresses to signal your position.

Wild animals pose a physical risk, particularly in dense forest where you can surprise, and be surprised by, an animal at close range. Elephants, buffaloes and other large mammals can attack and kill you, particularly if they have young or if people hunt in the area. Respect large cats, which can prey on people. Hippos attack boats and trample people. Constrictors (anacondas and pythons) can also prey on people. Be aware of venomous snakes and know what to do if you are bitten. Rehabilitant animals can be particularly dangerous, as they have little fear of humans. Caiman and crocodiles also pose risks. Check if it's safe before entering water.

Use a bed-net if there's a risk of biting insects. Use insecticides and repellents cautiously; they can be toxic to you and your environment. Food and scent, including perfume and soap, can attract unwanted attention from animals.

Other risks, with levels of severity ranging from irritating to lethal, include bites from dogs and other animals (with the risk of rabies infection), bee and wasp stings (take an epinephrine autoinjector in case you develop an allergy), ticks (check your skin for ticks, brush off those that are unattached, remove embedded ticks carefully with a tick remover or sharp pointed tweezers to avoid infection, and watch out for subsequent rashes or flu-like symptoms that indicate bacterial infection), leeches (use leech socks, tuck your shirt in), spiders and scorpions (shake out your clothing, bedding, and footwear before use, check where you put your hands), ants, flash floods, tropical storms with high winds, tree falls, and fires. Check the local situation, take advice, listen to it, and take precautions. Your study animals may alert you to danger; if they show behaviours that indicate danger, check your own safety.

If you're staying at an established field station, listen to the safety induction and make sure you understand. Respect the rules. If there isn't a safety induction, ask for one. Good practice includes keeping a log (a board or a book) of where people have gone when they leave camp and when they are expected back and following an agreed upon procedure if someone goes missing. If you are at a small camp, agree a safety protocol and follow it. Always be aware of your surroundings. A comfortable field station can lull you into a false sense of security.

Health and safety are a particular concern if you work very closely with animals, for example at an animal sanctuary. Animals grab, scratch, and bite. Get advice on how to behave and follow it. Not all institutions have high health and safety standards, so get training somewhere that does.

20.9 PHYSICAL HEALTH

If you spend any length of time in the field, the chances are you'll get ill or have an accident. Of course, this can happen anywhere, but fieldwork is often in remote places with limited access to quality medical care. Diarrhoea is almost unavoidable, so understand oral rehydration. Other common concerns are respiratory problems, dermatological infections, and fever. Animals, blood, food, insects, and water can all transmit pathogens.

The field environment can be punishing. Take time to rest. Don't overexert yourself or push the limits of your abilities. Exhaustion is dangerous and disorientating. Beware of sunburn, heatstroke, and heat exhaustion. Wear sunscreen and a hat, if appropriate. Avoid the temptation to reduce the amount you drink to carry less water. This can lead to bladder infections and exhaustion.

Don't take unnecessary risks with your health. Don't copy field-workers who do so; choose your own risk levels. Don't imitate friends and colleagues from the local community who may have different infection histories to you.

Blowflies go by many names (mango fly, tumbu fly, putzi fly, etc.) and lay their eggs in clothes left out to dry. When the larvae hatch, they burrow into the wearer's skin and feed on subcutaneous tissue, causing a boil with an opening in the centre. Ironing clothes kills the eggs but is not always practical. You can remove the larva by covering the opening in the boil with petroleum jelly. This forces the larva to surface, then you can squeeze the boil out. Don't burst the larva and beware of infection.

Many diseases and pathogens carried by humans can be passed on to our study species, including diarrhoea and respiratory diseases, so don't work if you are ill, and ensure that you and your staff are vaccinated against transmissible diseases (Section 2.4).

Pay attention to hygiene. Wear gloves when collecting samples.

Learn as much as you can about health risks at your field site before you go to the field. Consult experienced colleagues, discuss your plans with your doctor (physician), and get all the immunisations you need. Plan this well in advance – some vaccines need to be administered several weeks in advance of departure or need several doses. You may need some vaccines, like yellow fever, to enter a country and vaccinations may not be available near your field site.

Pack prescription medication in your carry-on bag, so that you have it if your luggage is delayed or lost. Check that non-prescription drugs are legal in the country you are going to or travelling through.

Consider the implications of any allergies and existing chronic health conditions you have for fieldwork. Such conditions can also develop in the field and may be exacerbated by physical or psychological stress. Take sufficient medication with you for your field season. Discuss with your doctor (physician) the potential consequences of interrupted treatment, potential interactions between your condition, any treatment and the drugs you may take as part of your fieldwork (e.g. malaria prophylaxis), and health issues particular to your field site. For example, if you are HIV positive, live/bacterial vaccines may be inappropriate, or you may need more vaccines than seronegative researchers, depending on your treatment history. If you are 'undetectable', you should be able to tolerate most vaccines, but always consult your doctor prior to vaccination. Some countries require an HIV test for entry and may bar entry either completely or change the duration of your stay if you are HIV positive, but these laws change frequently (www.hivtravel.org

provides updated lists). Bear in mind local stigma before disclosing your status and remember that showing your medication to someone may reveal your status.

Get first-aid training (wilderness first-aid training courses are available in some countries); pack a medical kit and know how to use it. Find out where to get medical advice when in the field in case you need it. Keep a copy of emergency numbers like the nearest hospital or clinic. You may need to take sterile needles with you because dirty needles are repackaged and resold as clean in some places. If you need treatment in a hospital, you may need to buy supplies like needles and sutures from the pharmacy.

Untreated water can harbour pathogens from human and animal faeces that cause gastrointestinal illness or fatal diseases, including bacteria like cholera, typhoid, *Escherichia coli* and *Salmonella*, protozoa like *Entamoeba*, *Giardia*, *Cryptosporidium*, and viruses like polio and hepatitis A. Boiling water for at least 1 min (longer at altitudes greater than 2000 m) kills all pathogens but uses up fuel. Microfiltration removes most bacteria and protozoa, but not viruses. Disinfection with iodine or chlorine is effective for some protozoa (but not *Cryptosporidium*), bacteria, and viruses. Purification tablets make water taste unpleasant, but neutralising tablets help. Ultraviolet light water purifiers kill some microorganisms but only work with clear, particulate-free water, so you need to filter the water first. They use rechargeable batteries. Consider your water treatment system carefully and follow the manufacturer's instructions for filters, chemical disinfection, and purifiers.

Take malaria very seriously. Seek advice on prophylaxis and choose carefully. Some doctors will strongly recommend mefloquine (sold under brand names, so check the generic name). Some people take this with no problems, but it can have serious side effects and induce mood disorders and hallucinations. This can be a particular problem for fieldworkers who are already subject to the stresses of a new environment, and who may be working alone and at night. Avoid mosquito bites and take your prophylaxis. Understand the symptoms of malaria, never ignore them, and seek diagnosis if you detect them. Take, and use, rapid-testing kits and emergency medication. These are not a replacement for seeking professional help.

Find out what other infectious diseases occur at your study site. There are no prophylaxes for dengue fever or zika virus, both of which are carried by mosquitoes. Zika poses risks to people who might become pregnant. Current understanding suggests that the virus can be transmitted sexually for at least 6 months after initial infection. Black flies along rivers transmit onchocerciasis. Schistosomiasis (bilharzia) is common in rural agricultural areas of the tropics.

Take the same precautions against sexually transmitted diseases that you would take anywhere else. If you use condoms, you may need to bring them with you as they may not be readily available, or it may not be culturally acceptable to purchase them. HIV may be prevalent where you work. If you use pre-exposure prophylaxis (PrEP) to protect against HIV seroconversion, bring enough to take the required daily dose. If you think you've been exposed to HIV, post-exposure prophylaxis (PEP) may be available at local hospitals or clinics but must be taken within 72 h of exposure to be effective.

If you wear corrective spectacles or contact lenses, take spares and supplies with you. If you cannot see well without them, include this in your risk assessment. For example, consider how you would cope if you were chased by an elephant and lost your spectacles or lenses in a swamp.

If you menstruate, take sanitary supplies with you because you may not be able to find these in the field. Consider how you will dispose of sanitary items in the field and how to guarantee cleanliness. Non-applicator tampons are easier to pack and to dispose of than pads. Menstrual cups alleviate disposal concerns entirely, but you need to sterilise the cup. If you haven't used a cup before, try one before you leave for the field to make sure you're comfortable using it. Remember painkillers and a hot water bottle if you need them for cramps. Field-work can influence your cycle; you might skip a cycle, or you might bleed for longer than usual. If you use contraceptive pills to regulate your menstruation, remember that diarrhoea can eliminate the hormones, so you'll menstruate.

If you use contraception (birth control), talk to your doctor about practicalities, and don't change method just before you leave. Remember that diarrhoea and some medications interfere with the contraceptive (birth control) pill. Take advice on whether these include your malaria prophylaxis.

20.10 MENTAL HEALTH

Fieldwork can be exhilarating. However, it can also be shockingly lonely, extremely uncomfortable, and exhausting. Events can take a heavy emotional toll on you. These may be the sorts of event that happen anywhere (e.g. a relationship breakup, a setback in your research, a death), but you may feel extremely isolated in the field. Depression and anxiety are common among fieldworkers. You may be particularly vulnerable if you have a pre-existing condition, but these issues can affect anyone.

As a fieldworker, you may be away from home for long periods, and you may be completely out of contact with people you are used to communicating with regularly. You may miss important events at home. Some people are comfortable with this; others are not. If communication is important to you, then budget for it. Your field site may have an Internet connection. If there is mobile (cell) phone coverage, make sure your phone is unlocked so that it can be used on any network and buy a local SIM card to make calls. You might need to be inventive – for example, by hoisting your phone up a tree to get a connection.

Some people find keeping a diary (journaling) a very effective way to cope with the difficulties of fieldwork. Also make friends in towns close to your field site and visit them. Plan, and take, breaks to relax. Communicate your feelings with your team, and pay attention to their mental health as well as your own.

If you or your family have a history of mental health issues, avoid mefloquine for malaria prophylaxis, as it can trigger or dangerously exacerbate these issues.

20.11 RETURNING HOME AND RE-INSERTION SYNDROME

Many fieldworkers bear lasting consequences of parasite infection. If you're conducting international fieldwork, consider having a full parasite check before you leave your host country. Medical and laboratory staff will be familiar with the pathogens you may have been exposed to. Many parasites can be treated quickly and effectively, but try not to collect them all the same, as they can have lasting effects.

If you are unwell when you return from the field, particularly if you have symptoms of malaria or flu-like symptoms, tell your doctor where you have been and what the possibilities of infection are. Tests for some infections are unreliable.

Returning from the field can be psychologically and emotionally very challenging. For students, this reverse culture shock, or **re-insertion syndrome**, may be compounded by the challenges of moving from the familiarity of data collection to data analysis.

Reverse culture shock happens because your concept of home has changed. Your original home may have changed while you were away, and you have adapted to another home. The more comfortable you were in the field, the harder the transition is.

Experiences vary, but you may feel like no one wants to hear about your life in the field, or that no one understands it. This feeling of alienation can be compounded by missing your friends and your routine in the field. You may be shocked by the pace of life, the materialism, or the abundance of choice in the supermarket when you return from the field.

Just like in the field, you can manage your feelings by talking with close friends, journaling, and talking with colleagues who have been through similar situations. You will adjust, with time.

Keep in contact with your friends in the field. Social media is great for this. One of the great advantages of fieldwork is that you get to know people you would not otherwise meet. The disadvantage is that you may not see them as often as you would like.

20.12 CHAPTER SUMMARY

In this chapter, we've seen that:

- Fieldwork is exhilarating and life-changing but can be extremely challenging.
- Preparation is crucial, but we can never be fully prepared, so we must also be flexible, patient, and determined.
- Challenges include obtaining permissions, coping with logistics, mitigating risks to personal safety, navigating the social context, prejudice, natural hazards, and maintaining physical and mental health under extremely challenging conditions.
- We must consider these challenges and risks carefully and consult experts in the local area in advance.
- We need an exit strategy in case of emergency.
- We may encounter shocking poverty in the field and wish to help. Doing so requires consultation with those we wish to help and careful reflection. There are no simple solutions and we are not saviours
- Our conduct will influence the experience of future researchers. We are ambassadors for primates and for primatology.
- Returning from the field can be psychologically and emotionally very challenging.

Once you have collected your data, in the field or elsewhere, you can put your detailed analysis plan (Chapter 15) into practice. In the next chapter, we look at analysing and interpreting data.

20.13 FURTHER READING

Clancy KBH, Nelson RG, Rutherford JN, Hinde K. 2014. Survey of academic field experiences (SAFE): Trainees report harassment and assault. *PLOS ONE* 9: e102172. https://doi.org/10.1371/journal.pone.0102172. Reports on a survey of experiences of sexual harassment and sexual assault in fieldworkers showing that the primary targets are women trainees, and that perpetrators are predominantly senior to them professionally. Suggests policies to improve the situation.

Ice GH, Dufour DL, Stevens NJ. 2014. *Disasters in Field Research: Preparing for and Coping with Unexpected Events*. Lanham, MD: AltaMira Press. Provides a wealth of practical suggestions to avoid, or at least minimise, the impact of the unexpected, with real-life examples from researchers in a variety of disciplines.

Jolly A. 2016. *Thank You, Madagascar: The Conservation Diaries of Alison Jolly*. London: Zed Books. A beautifully written account of the late Alison Jolly's experiences in Madagascar.

Morgan BJ. 2012. Notes from the field: A primatologist's point of view. *Nature Education Knowledge* 3: 8. Describes daily life as a field primatologist in Central Africa.

Smith DS. 2012. Travel medicine and vaccines for HIV-infected travellers. *Topics in Antiviral Medicine* 20: 111–115. A summary of considerations for vaccination in immunocompromised people.

Sohn E. 2019. Ways to juggle fieldwork with kids in tow. *Nature* 570: 405–407. https://doi:10.1038/d41586-019-01909-w. Advice on conducting fieldwork with children.

Werner D, Thuman C, Maxwell J. 2015. *Where There Is No Doctor: A Village Health Care Handbook, Revised Edition*. Berkeley, CA: Hesperian Health Guides. A classic healthcare handbook on how to prevent, recognise and treat many common sicknesses. Search online for free pdf versions in multiple languages.

Also see ethical travel websites for advice.

21

Analysing and Interpreting Data

Once we have collected our data, data analysis and interpretation allow us to test our predictions and interpret the results. This can be daunting because it's a big change from data collection. It's very unlikely that we will have collected exactly the data we set out to collect, but our analysis plan (Section 15.10) will keep us on track and avoid the questionable research practices associated with aimlessly exploring our dataset (Section 3.5).

In this chapter, I begin with data organisation and initial data analysis, go on to hypothesis testing, calculating effect sizes and confidence intervals, and end with interpreting results.

21.1 ORGANISING YOUR WORK

Organise your data in the spreadsheet you prepared before you started collecting data (Section 15.11). If your data are not already in electronic format, this will avoid your spending a great deal of time entering data you don't need. It's unlikely that you need to transcribe all your raw data; just extract the summary values you need for your analysis.

Keep your work organised. This takes time but will save time when you need to re-run analyses. Name your data files carefully and back them up. Storage media fail, so plan for this. Keep a copy of your raw data and keep a clear record of everything you do to it, including all the decisions you make. You'll need this when you write up your report and when you revise it. An easy way to do this is to use R and save your analysis code.

Document, organise, and store your data carefully to ensure they can be used in future. If you don't keep careful records, you will forget what the variable names refer to, what codes mean, the decisions you took, and which version of a file to use, making your data unusable. Store your data, with all the information needed to understand it, in a permanent format. Annotate your analysis code so that someone else can understand it (that other person might be the future you) and save it, so that you can reproduce your own results easily. If you use a commercial statistical software package, make sure you save your data and results in a format that you can open when your licence expires.

21.2 STICKING TO AN ANALYSIS PLAN

No project runs to plan, so you'll need to revise your original analysis plan, but always focus on your predictions and always prepare a detailed analysis plan *before* running your analyses. To avoid questionable research practices associated with data analysis (Section 3.5):

1. Follow a pre-defined analysis plan. If you have to deviate from it, note any changes you make to it and justify them.
2. Resist the temptation to import your data into a statistical programme and run multiple analyses on the same dataset, with different combinations of variables, until you find a result that you find interesting, then report that as though that result came from an analysis you intended to do from the start.
3. Seek converging evidence. If two or more measures show the same result, then a pattern may be robust. Similarly, you should be able to detect it using multiple methods, with and without outliers (Box 5.1), covariates, and so on.
4. Never choose a one-tailed test (Section 15.2.2) after examining the direction of a relationship. In other words, don't succumb to the temptation to replace a two-tailed test that yields a p value of 0.05–0.10 in the predicted direction with a one-tailed test to reduce the p value to <0.05.
5. Use family-wise error tests to correct for an excess of false positives in planned multiple comparisons. The Bonferroni correction is popular but greatly reduces statistical power. The Benjamini–Hochberg procedure is based on the **false discovery rate** – the fraction of statistically significant results that are false positives – and usually provides better statistical power than the Bonferroni correction.
6. Don't arbitrarily split continuous variables into discrete variables without very good reason. Doing so reduces the information in

your dataset and thus reduces your statistical power. If you do have good reason to split your data into groups, choose meaningful cut-offs and define them prior to analysis. Don't try out multiple cut-offs to maximise your chances of finding a statistically significant result.

7.   Tell the full story, not a selective account. Describe and justify all the analysis decisions you make and all analyses you run when reporting our results.

8.   Be honest when you look for patterns in data, then hypothesise after the results are known (Section 3.5). State that your findings emerged from the data, not from theory, and that the hypothesis needs to be tested with an independent dataset.

9.   If you run exploratory analysis, split your data into two random subsets. Use one to construct your hypotheses, and then test the hypotheses on the other subset. Of course, you lose statistical power if you do this (Chapter 16).

10. Use statistical significance as just one piece of evidence when assessing findings, in addition to existing knowledge, plausibility, quality of the study design, and data (Section 5.10).

11. Make the data and code (e.g. R scripts) underlying your results publicly available when reporting your results, so readers can assess them.

21.3 PLOTTING DATA TO CHECK FOR ERRORS AND CHECKING THE ASSUMPTIONS OF YOUR MODELS

Begin by checking your data for completeness and accuracy. Inspect your data visually. Check the shape of the distribution for each variable. Check for outliers, skew (which biases the mean), and pointiness (**kurtosis**, which biases the spread).

Examine the relationship between variables in bivariate plots. Use box-plots to examine differences between groups. Check for floor and ceiling effects, which will affect your analyses (Section 14.5).

Don't use this step to generate hypotheses or you will introduce researcher degrees of freedom (Section 3.5). Follow your analysis plan. One option is to ask someone who is blind to your predictions to assess the distribution of your data.

Check that the assumptions underlying your models (Chapter 15) hold before looking at the results. If you look at the results first, and they are exciting, you will be tempted to explain away problems with the assumptions.

Remember that unless you use a test that explicitly accounts for dependence in your data, your data should be independent (Section 15.3).

Statistical packages provide statistical tests of the assumptions underlying models, but you should also inspect the plots your statistical software produces because you may still be able to use a model, even if the test is significant.

If your data don't fit the assumptions of your models, explore any problems. Investigate outliers (Box 5.1). If they are simple errors, you can correct them. If they are not, then you can't simply ignore them.

If your data don't fit the assumptions of parametric models, you may still be able to use these models, because some tests are robust to such problems. You can transform data (Section 15.2.1) to improve the fit, but don't experiment with different transformations in the hope of getting a significant result from your model. Also consider resampling methods (Section 15.2.1).

Check for collinearity among predictor variables. Use statistical software to calculate a **variance inflation factor** (VIF) for each predictor variable. A value of 1 indicates no collinearity. Arbitrary rules of thumb propose a VIF of 5 or 10 for high collinearity but use your own statistical judgement rather than arbitrary rules to evaluate your results. Standardising your variables (subtracting the mean from all observed values of that variable and dividing by the standard deviation) can help to reduce VIFs.

When using non-parametric tests with small sample sizes (typical in primatology), use exact tests, not asymptotic tests. Your statistical package may not do this for you automatically.

21.4 RUNNING ANALYSES

Once you have checked all the assumptions of your tests, you can run your analyses. It can take time to fit an appropriate model. Don't just press buttons in statistics packages to see what answers you get. You will get answers, but they are likely to be nonsense.

Once you have results, evaluate their stability and the influence of single data points, if possible, using replication or cross-validation. Assess the relative risk of false positives and false negatives when considering your results (Section 5.6), but don't do *post hoc* power analyses (Box 16.1).

21.5 CALCULATING EFFECT SIZES AND CONFIDENCE INTERVALS

Assess and report effect sizes and confidence intervals as well as, or instead of, *p* values (Chapter 5). In other words, report how much of an effect you have detected, and the precision with which you have measured that effect, rather than only whether the effect is statistically significant.

Show important data in a figure, so that readers can evaluate the effect for themselves.

If you transformed your data for analysis, transform your descriptive data and effect sizes back to the original scale of measurement when reporting them. These back-transformed values are not always the same as those calculated using the untransformed data.

21.6 INTERPRETING THE RESULTS

Remind yourself of what you can and can't conclude from a significance test (Section 5.5). Statistical significance tells you about the improbability of your findings but does not tell you anything about the importance of a result. A statistically significant result may be trivial, and a statistically non-significant result may be important. Moreover, the use of *p* values can encourage dichotomous thinking, which is unlikely to be scientifically appropriate.

You cannot conclude that a highly significant result is more important than one that is closer to the cut-off, although researchers often do. Similarly, you cannot conclude that a non-significant result means that there is no effect, although, again, researchers often do. A non-significant result is merely inconclusive. There may be no effect, or your study may lack the power to detect it. A further common error is to conclude that groups are the same, because a statistical test comparing them is non-significant. However, you can't conclude this, either (Section 5.5).

Challenge yourself to describe the effect you find as clearly as possible in real-world terms, in addition to the statistical significance.

If your sample is representative, then your results hold for the population you studied, at the time you studied them, under the conditions of your study, and within the observed range of the variables you measured. Generalising beyond these bounds is speculation.

Statistical models are valid for the range of values in the dataset used to fit the model. We can use them to predict values of the outcome model within this range (**interpolation**), but we must be cautious when predicting beyond the range of the observed values (**extrapolation**), as we don't have evidence to support this speculation. For example, we might model a relationship between age and mass for juveniles but applying this model to adults would make no sense.

Examine your results carefully, taking into account your understanding of your study animals. Are any of your results counterintuitive? If they are, why might this be? Could any relationship you detect might be the result of a confounding variable (Section 5.3)?

If you have a statistically significant interaction in your model, you cannot interpret the main effects of the corresponding variables without considering the interaction. For example, if the relationship between body mass and dominance rank depends on the sex of the animal, then simply examining the relationship between mass and rank, without splitting the data by sex, would lead us to draw erroneous conclusions.

You can now determine whether your findings support your hypothesis. If they do, you can retain the hypothesis for the moment, although subsequent studies may refute it. Remember, our findings can support a hypothesis, and exclude alternative explanations, but we can never prove a hypothesis (Section 1.1). You can now propose further tests to rule out possible confounds and alternative explanations.

If your results don't support your hypothesis, then you may need to refine it, or to abandon it. Of course, you'll be disappointed if you find no support for a favoured hypothesis, but science progresses by ruling out explanations and good scientists will appreciate a carefully designed study, regardless of the outcome. Don't allow wishful thinking to influence your interpretation.

However carefully we design a study, we learn as we conduct it. Lack of support for a prediction may mean that the study wasn't appropriate to test it. What we learn will help to design an improved study.

Integrate your results with those of earlier studies. Look for similarities and reflect on what might explain any differences. Consider combining your new data with a systematic review of current data and a meta-analysis of all the evidence to date (Section 11.8).

21.7 CHAPTER SUMMARY

In this chapter, we've seen that:

- Data analysis and interpretation are exciting but can be daunting.
- We will probably need further statistical advice at this stage.

- Data curation and storage are vital.
- It's very unlikely that we will have collected exactly the data we set out to collect, but we must avoid the temptation to explore our data aimlessly, looking for patterns, to avoid detecting and reporting false positives.
- We must also resist the temptation to run multiple analyses on the same dataset in the hope of finding support for a prediction. Instead we should conduct pre-planned analyses and specify any *post hoc* analyses clearly.
- We should test the assumptions underlying the tests we use and assess the relative risk of false positives and false negatives when considering our results.
- We should calculate effect sizes and confidence intervals as well as assessing their statistical significance.
- We interpret our results to draw conclusions about the hypotheses that we tested.

Once we have tested our predictions and interpreted our findings, we can write a report. In the next chapter we look at how to do this.

21.8 FURTHER READING

In addition to the readings I recommend in Chapters 5 and 15, try:

Benjamini Y, Hochberg Y. 1995. Controlling the false discovery rate: A practical and powerful approach to multiple testing. *Journal of the Royal Statistical Society, Series B* 57: 289–300. https://doi.org/10.2307/2346101. Proposes the false discovery rate and a procedure to control it.

Ihle M, Winney IS, Krysalli A, Croucher M. 2017. Striving for transparent and credible research: Practical guidelines for behavioral ecologists. *Behavioral Ecology* 28: 348–354. https://doi.org/10.1093/beheco/arx003. An example of the application of open (transparent) practices to a field.

Mundry R, Fischer J. 1998. Use of statistical programs for nonparametric tests of small samples often leads to incorrect P values: Examples from Animal Behaviour. *Animal Behaviour* 56: 256–259. https://doi.org/10.1006/anbe.1998.0756. Highlights the need for exact (rather than asymptotic) test procedures for non-parametric statistics.

Neuhauser M, Ruxton GD. 2009. Round your numbers in rank tests: exact and asymptotic inference and ties. *Behavioral Ecology and Sociobiology* 64: 297–303. https://doi.org/10.1007/s00265-009-0843-1. Warns against pseudo-precision in ranking data.

Nosek BA, Ebersole CR, DeHaven AC, Mellor DT. 2018. The preregistration revolution. *Proceedings of the National Academy of Sciences of the United States of America* 115: 2600–2606. https://doi.org/10.1073/pnas.1708274114. Introduces preregistration and addresses challenges to it, including how to handle changes to the methods and violations of assumptions.

Parker TH, Bowman SD, Nakagawa S, Gurevitch J, Mellor DT, Rosenblatt RP, DeHaven AC. 2018. *Tools for Transparency in Ecology and Evolution.* http://doi .org/10.17605/OSF.IO/G65CB. A checklist of questions to maximise transparency in science.

Pike N, 2011. Using false discovery rates for multiple comparisons in ecology and evolution. *Methods in Ecology and Evolution* 2: 278–282. https://doi.org/10.1111/ j.2041-210X.2010.00061.x. Discusses the advantages and disadvantages of analyses based on the false discovery rate and provides spreadsheet programmes to calculate false discovery rates.

22

Writing a Scientific Report

With our analyses done, we can write up our research project. This chapter builds on the general advice for writing in Chapter 6 and focusses on how to write a scientific report. I provide general guidance for writing your report, then cover each section of the manuscript in turn. I focus on primary research articles, because these are the main way in which we disseminate new research, but much of the advice applies more generally to theses and dissertations (Box 22.1). I also include information on how to write a good review article (Box 22.2), and reports to funders, stakeholders, authorities, and hosts (Box 22.3).

22.1 GENERAL GUIDANCE

Scientific reports have a standard format, with some variation. This should be familiar from your reading (Box 11.1).

Begin by writing out your study aims clearly and concisely. Your entire manuscript centres on these and stating them plainly will tell you exactly what goes in each section of your report. The introduction puts the aims in context, the methods explain how you addressed them, the results explain what you found and the discussion puts the findings into context and interprets them. The structure of a manuscript resembles an hourglass: the introduction begins with the broad context, then narrows progressively to the specific aim of your study. The methods and results concentrate on the specific aim (the neck of the hourglass). The discussion begins with the specific focus of the study, then broadens the focus to address the bigger picture, reflecting the introduction.

A table of hypotheses, predictions, and whether your study supports them can make a very useful framework for your manuscript.

Box 22.1: Theses and dissertations

Requirements for theses and dissertations vary by country and institution. Reports may constitute a monograph or a compilation of research articles related to a theme, usually with a general introduction to the entire project at the beginning and a general discussion of the contribution you have made at the end.

A compilation avoids the need to rewrite thesis chapters for publication. Articles are often collaborative, with the student as first author, and you may need to state which aspects of the work are yours in your thesis submission.

A thesis has a similar structure to a journal article (Box 11.1), with the addition of a contents list (you can make these in word processing software): Title, Abstract, Acknowledgements, Contents, List of Tables, List of Figures, Introduction, Methods, Results, Discussion, References, Appendices.

A longer thesis with several results chapters takes the format Title, Abstract, Acknowledgements, Contents, General Introduction, General Methods, Results 1, Results 2, Results 3, etc., General Discussion, References, Appendices. The General Introduction is the overall rationale for the project and ends with the major aims and structure of the thesis. The General Methods chapter covers methods that apply to the study as a whole. The results chapters each have their own introduction, methods, results, and discussion, specific to the aims of each chapter. The General Discussion draws out the major themes of the project.

The contents list is a list of chapters and sections, and separate lists of tables, figures, and abbreviations. Number the chapters, headings, and subheadings for ease of reference. Include tables and figures in the text, following the paragraph in which you refer to them. Number them with the chapter number, then sequentially within the chapter (e.g. tables in Chapter 3 are: Table 3.1, Table 3.2; figures in Chapter 4 are: Figure 4.1, Figure 4.2).

Once you have clear aims, you can plan your writing. Many people begin by writing the methods, followed by the results, before moving onto the discussion and introduction. See what works best for you.

You already have material from your proposal, so there is no need to start with a blank page (unless that works well for you). However, you can't just copy and paste material from your proposal; you need to

---

Box 22.2: Writing a good review article

Review articles summarise and synthesise what we know and provide a framework to guide future work. Reviews are very useful preparation for a research project (Section 11.8), or a grant proposal (Chapter 18).

The synthesis aspect of a review is key – reviews are not as simple as a summary of what existing studies of a topic show (Box 11.3). Instead, reviews add value by setting an agenda for future work.

A review can provide new theoretical understanding of a topic. This might be through identifying common patterns, highlighting untested assumptions or other gaps in our knowledge, clarifying our thinking about a topic, redirecting research efforts, or posing new questions.

A review might also connect two different fields, to show how applying theory or methods from one field to another field would provide new insight into existing questions or generate new questions.

A review can also evaluate different approaches to a problem, examining the advantages and disadvantages of each, discussing limitations, highlighting common problems, and proposing solutions.

Review articles have a more flexible structure than an empirical report, but they still need a clear structure. A diagram that represents the relationships among ideas (a **concept map**) is very useful to get your thoughts in order before you draft a review. You can include this in your review to guide your reader.

As with empirical reports, the structure of a review centres on the aim, but in the case of a review, the evidence you provide to underpin your argument is from published articles, rather than your own data. Begin with a broad overview of the context, then state the aim of the review. Outline the structure of your article to your readers. Give your conclusion early in the review, to avoid taking your readers on a mystery tour, then support your conclusion with evidence from the literature. End with future directions.

Provide a fair and balanced evaluation of the literature. Briefly explain the criteria you used to choose the studies you include (These should not be only from your research group.) Include evidence both for and against your line of reasoning. Identifying and counteracting any objections strengthens your argument.

Box 22.2: (cont.)

Imagine what sceptical readers might think and anticipate and respond to their concerns.

Systematic reviews are exhaustive, using stated criteria to identify and synthesise all the literature on a topic. These are rare in primatology, although they are fairer than selective reviews, and form the basis for meta-analyses (Section 11.8). In a selective review, include older, seminal work that provides the foundations for current research and the most recent publications on a topic, to describe the present state of understanding of a problem. Check for older studies that may have been overlooked. Don't ignore work that you can't find easily online or rely on secondary sources; which may be wrong.

Include only relevant aspects of the studies you refer to. Ask yourself whether the information you provide is necessary to underpin your argument. Unnecessary details distract your readers. The amount of information to provide about individual studies depends on the focus of the review. For example, is the exact sample size in a study relevant to your argument? Perhaps all the readers need to know is whether it is particularly large or small. If you're writing for a non-primatological audience, you may not need to give species names, but for primatologists, you probably do.

If you criticise earlier work, remember that we learn from earlier studies. Never criticise individual researchers, only the work.

FURTHER READING

Pautasso M. 2013. Ten simple rules for writing a literature review. *PLOS Computational Biology* 9: e1003149. https://doi.org/10.1371/journal.pcbi .1003149. On why we need reviews, and useful tips on how to write them.
Sayer EJ. 2018. The anatomy of an excellent review paper. *Functional Ecology* 32: 2278–2281. https://doi.org/10.1111/1365-2435.13207x. A very useful editorial on how to write a review.

revise it. The audience for a report is very different to that for a proposal and the writing style differs. Of course, you also now have results to report and interpret.

Remember that your aim when writing a scientific report is to convey information to your readers, allowing them to check and

Box 22.3: Reports to funders, stakeholders, authorities, and hosts

In addition to research articles and dissertations, we may need to write other types of reports, including interim progress and final reports for our funders and reports to hosts, stakeholders, and other interested groups or communities.

As with all writing, our aim is to communicate information. In each case, we should consider the audience and what they want to know from us and explain our research in a suitable format and language.

Reports to Funding Bodies
Check when you need to provide reports to your funders, and what they expect. If the funder gives instructions, follow them. If they don't, provide a brief summary of your project aims, then describe what you did, explain any changes to the project (you should inform funders about major changes before making them), how you spent the funding, the results (preliminary findings are fine), your conclusions, and future plans, including dissemination. Include photos, particularly if the report will be published (e.g. in a newsletter). Find out whether you need to include a financial report, or receipts, and follow any instructions for submitting these.

Reports to Stakeholders, Authorities, and Hosts
Influencing policy and practice requires a two-way exchange of ideas and experience and co-produced research which involves partners from the beginning (Section 9.4). As part of this, you may need (or wish) to provide reports to stakeholders, as well as to relevant authorities and hosts. If you do, find out what your audience needs to know, and how they would like the information to be provided. Ask yourself what you want them to do with the information.

Begin with a short overview that communicates your key messages to readers who may not read the whole document. Then provide an introduction to your study, brief methods, findings, interpretation, and recommendations (if relevant). Highlight the most important information and convey it clearly, using diagrams and clear graphs. Don't just copy material you use in scientific reports; this is a different audience. Provide accurate information. Make specific and realistic recommendations, if appropriate,

> **Box 22.3:** (cont.)
>
> support these with evidence from your findings and define priorities. Share a draft report with your partners and seek feedback as to whether it is useful, then revise it to reflect this feedback.

interpret your results for themselves. To achieve this, write clearly, simply, precisely, and concisely (Chapter 6). Avoid confusing or distracting your readers.

If you are preparing a manuscript for a specific journal, read its Instructions for Authors before you start. Each journal has its own requirements. As an editor, my most common request for revisions is *Please comply with the Instructions for Authors*.

You will probably have ideas about other sections or for future research as you write. For example, as you write your results, you may well have ideas for the discussion. Write these ideas in a notepad or in a separate document and keep writing the section you're working on. Expect to move text around.

22.2 THE TITLE: THE CORE MESSAGE OF YOUR REPORT

The title is the first part of your report that potential readers come across. Potential readers will decide whether to read further based on your title.

Your title should reflect the major point(s) of your report. It should be interesting, clear, concise, informative, unambiguous, and accurate.

Place the most important part of your title first. Any information that appears after a colon may not appear in a list of online search results. Opinions differ on informal two-part titles and the use of direct questions, so check the Instructions for Authors. If well done, informal titles are memorable. If not well done, they puzzle or annoy the reader. Search engines may prioritise the first part of a title, making an informal title a disadvantage if it doesn't contain your key terms.

Titles can be declarative, with the answer to a research problem (e.g. *Male mandrills prefer to mate with high-ranking females*), pose a question (e.g. *Do male mandrills prefer to mate with high-ranking females?*) or descriptive (e.g. *Mating strategies in male mandrills*).

Remove unnecessary words and phrases (Box 6.3), but don't stack modifiers (adjectives or nouns which modify a noun) before a noun (Box 6.4). In other words, *wild mandrill* is fine, but *wild, adult, high-ranking mandrill* is not.

Choose an appropriate level of generalisation. For a primate journal, you probably need to include the study species name. For a more general journal, you can be more general (e.g. *male mate choice in a polygynandrous primate*) but beware of over-generalisation.

An excellent example of a title is *Primate males go where the females are*.[1] Looking for this, I hadn't remembered the author, but the phrase stuck in my mind. We rarely remember very long titles. (Do you know the whole title of the book we refer to as *The Origin*?)

Avoid linking papers together in a series because this can lead to complications when they move through the review process at different speeds. Moreover, if the titles are similar, then the papers will be confused once published. Instead, use the introduction to make the link between your publications.

Journals often ask you to provide a short version of the title for use at the top of each page. Link this running title clearly to the full title.

22.3 THE INTRODUCTION: WHY YOU DID THE STUDY

The introduction gives the context to your study. It reviews what we already know about the study question and system, explains what we do not yet know (i.e. the gap your study fills), and provides the detail your readers need to know to put your case study in context. It should begin broadly with established knowledge about a question (not a species), narrow the focus to the specific question that you address, and then introduce what you did to answer that question.

Introductions vary in length from two to three paragraphs to a more extensive review. If you have written a literature review (e.g. to identify a specific research question) then you can turn it into an introduction, but you cannot just paste it in. An introduction should only include the information relevant to the immediate study subject and the reasons for doing the research. There is no need to review the entire general context to your study.

The introduction has three parts: the general introduction, the specific introduction, and the study set-up. Include these as headings in

[1] Altmann J 1990. *Animal Behaviour* 39: 193–195.

your draft, but take them out in your final report. Authors often muddle these sections, introducing their case study in the general introduction, presenting further general information in the specific introduction, or including multiple (and different) statements of their aim in different places in the introduction.

If your aims changed during your study – and they usually do – write the introduction to the project you conducted, not the project you set out to conduct.

22.3.1 The General Introduction: Your Question

Open your introduction with a broad statement about the theoretical context of your study. Then present the general research question that you will address. Review the literature on the subject, highlighting why the question is important, summarising our current understanding, and clarifying what we do not yet know. This section should be framed conceptually and broadly. This is where you develop your ideas before you mention the specific case study, species, or region that you address. In other words, do not begin with your study organism.

Focus on theory, not the history or the number of papers published on a topic. Highlight the major questions and hypotheses rather than describing the chronology of ideas. Don't write *x did this in the 1970s, then y did this in the 1980s*. If there are few papers on a topic, that may indicate that it is of no interest rather than a need for more work. Beware of providing more than a couple of citations to support a claim that there are few studies of a topic, or of using only one or two citations when claiming that there are many studies of a topic (unless you cite reviews).

Summarise and synthesise the existing literature. Don't list articles without providing information about their content. This isn't helpful to your reader. Be precise. Review the findings of a study, rather than the general topic they cover. Don't provide one paragraph per article that you have read; such paragraphs often look like they are derived from the abstracts. Instead synthesise what the articles show, including only the information relevant to your report.

Present what we know concisely and cite the most important work on the topic. Review the literature fairly (Section 3.2). Cite the earliest, most comprehensive, and highest quality work. Don't neglect the first authors to investigate a topic in favour of the latest replication or extension. Cite the original author of a hypothesis, not the empirical paper that you happened to read that tested it or cited it. Do not cite only work by your group when other groups also work on the same topic.

Critique earlier work fairly. Remember that science builds on earlier work, so don't highlight only the limitations of earlier studies and don't criticise them for addressing their own aims other than yours. Don't make sweeping statements about accepted ideas that you intend to refute (strawman arguments, Section 6.13). Journalism often presents science as overturning accepted ideas, but science usually proceeds by building on existing knowledge.

If you review multiple hypotheses, then provide signposts for your readers before reviewing the support for each one. For example, mate choice can be for direct or indirect benefits (the set-up). Direct benefits include food and offspring care (paragraph 1, with examples); indirect benefits include high-quality immune genes (paragraph 2, with examples).

By the end of the general introduction, your readers should understand what we know and what we don't yet know about a research question. This may include the need for replication (Box 1.2).

22.3.2 The Specific Introduction: Your Case Study

Once you have reviewed the general context, introduce your case study (e.g. your study taxon), and explain why it is a suitable choice to advance our knowledge of the question. You don't need to report everything we know about the system, but do report what the readers need to know to put your study into context.

Avoid claims of novelty or priority (e.g. *we provide the first evidence* or *we are the first to show*). All studies present new findings, so such claims are not necessary. The novelty of your study should be clear from your description.

No further general context should appear in your specific introduction. In other words, refer only to your model system from here.

22.3.3 The Study Set-Up

End your introduction with a clear, concise statement of your aim(s), the rationale for your study, and your hypotheses and predictions. Describe how your study design addresses the gap in our understanding that you identified, and allows you to test your predictions. There should be no surprises for readers here. The general and specific introduction should have provided all the relevant background, and no new topics should appear at this stage.

This section often begins with *The current study* ..., but you can simplify this to *We aimed to* .... In some reports, the introduction ends with a summary of the findings and their significance. I prefer to leave that to the discussion. Check any instructions you have.

22.4 THE METHODS: HOW YOU DID THE STUDY

The methods section is essential for your readers to assess your results and conclusions. Methods are different from methodology, which is the study of methods.

Use the past tense to describe what you did. Describe clearly how you carried out your study, including a description of your study site, details of the study subjects, and data collection (including dates), laboratory analysis, and statistical analysis – as appropriate – and in that order. Provide details of how you collected all data reported in the results but do not include methods for which you do not report findings. Justify your choice of methods, state advantages, and any limitations. Include the rationale and method you used to select your sample size. Define all terms carefully.

The methods should be appropriate, scientifically sound, current, and described in enough detail that the work could be repeated by another scientist with basic knowledge. Some journals move these details to **electronic supplementary materials**. Theses may contain a general methods chapter, in which case the methods section of a thesis chapter may refer to the general methods chapter.

Use subheadings to organise the content. Guide your reader by explaining briefly how your methods relate to your aims. For example, *To address aim 1, we measured* .... Describe all processes in order, to avoid confusing your readers (i.e. use *next* and *then*, not *after*).

Do not include results in the methods, excepting preliminary results that you used to design your study.

Describe your statistical analyses in a subsection entitled *data analysis*. Include details of data selection, data manipulation, and all data analyses conducted as part of a study, such that analyses can be reproduced and fully understood. Sharing your code is an excellent way to do this (Box 22.4). Include how you summarised data and report variability as well as the central tendency (Box 5.1). Describe how you tested your predictions in the same order as you introduced them. Justify the choice of all tests and address underlying assumptions. Name any programmes

Box 22.4: Good practice when reporting research

Good practice when reporting our research includes making our data and results publicly available (Section 3.5). Some funders and many journals now require open data and more encourage it. Archiving data in dedicated online digital **repositories** is more reliable than providing data as electronic supplementary material.

Benefits of Sharing Data
Sharing data benefits both the scientific community and the individual researcher by:

1.   Providing a stable back-up and encouraging careful data curation, minimising the risk of data loss. Data decay remarkably quickly. We quickly forget which data file we used for which analysis and what our codes mean, making the data unusable.
2.   Allowing others to check our analyses, reanalyse the data with improved models, and build on our work, increasing the overall contribution to science.
3.   Saving time when we get a request for data.
4.   Giving us credit if other researchers use the data.

Authors often worry when asked to share data, because they are concerned that other researchers may perform and publish an analysis before the original authors can do so. Making the data available sometime after publication (an **embargo period**) can mitigate such concerns. Moreover, we only need to share the dataset that underpins the article, not our entire database. In other words, we share only the data that we have already analysed.

Sensitive Data
In some circumstances, sharing data publicly has the potential to lead to harm, discrimination, or unwanted attention. This includes personal information about a person or people, and information that might have an adverse effect on conservation, such as the location of rare, endangered, or commercially valuable species. In such cases, we should refer to guidance on whether and how to publish or share such data safely, for example by allowing conditional access.

Box 22.4: (cont.)

Sharing Code

Sharing the code used to process and analyse data, as well as the data, makes the entire procedure transparent. It makes the analysis entirely reproducible, facilitates correction and reanalysis if software bugs are fixed, and promotes replication.

FURTHER READING

In addition to the reading in Chapter 3:

Australian National Data Service. 2017. *Guide to Publishing and Sharing Sensitive Data.* www.ands.org.au/guides/sensitivedata [Accessed 9 January 2019]. Best practice guidelines for sharing sensitive data.

Wilkinson MD, Dumontier M, Aalbersberg IJ, Appleton G, Axton M, Baak A, Blomberg N, Boiten JW, da Silva Santos LB, Bourne PE, Bouwman J, Brookes AJ, Clark T, Crosas M, Dillo I, Dumon O, Edmunds S, Evelo CT, Finkers R, Gonzalez-Beltran A, Gray AJ, Groth P, Goble C, Grethe JS, Heringa J, 't Hoen PA, Hooft R, Kuhn T, Kok R, Kok J, Lusher SJ, Martone ME, Mons A, Mons B, Packer AL, Persson B, Rocca-Serra P, Roos M, Sansone SA, Schultes E, Sengstag T, Slater T, Strawn G, Swertz MA, Thompson M, van der Lei J, van Mulligen E, van Schaik R, Velterop J, Waagmeester A, Wittenburg P, Wolstencroft K, Zhao J. 2016. The FAIR Guiding Principles for scientific data management and stewardship. *Scientific Data* 3: 160018. https://doi.org/10.1038/sdata.2016.18. Presents the rationale behind the FAIR guidelines, that data should be Findable, Accessible, Interoperable, and Reusable.

you use and state your α value (the probability of a false positive) and the β value (the probability of a false negative; Section 16.2).

Identify any ethical implications of the study and explain in detail how you addressed these. Specify guidelines and legal requirements and any licenses, permits, and consent acquired to carry out the work. Some journals require these in a separate subsection.

22.5 THE RESULTS: WHAT YOU FOUND

The results section reports your findings succinctly and in a logical sequence so that your reader can understand them. Use the past tense. Results are factual. Don't include further introductory material. Don't repeat the methods or include new methods, unless you conducted *post hoc* analysis (analysis informed by the data), in which case state this clearly (Section 3.5). Don't compare what you found to the literature or interpret what you found in the results. Leave that for the discussion. If you need citations, then that material almost certainly belongs elsewhere in your report.

Start with descriptive statistics for your key variables, then present the results of your statistical analyses. Begin each paragraph with the main finding of each analysis. It is very easy to lose your readers in a mass of descriptive and statistical results if you do not stick carefully to those that address your predictions. Use the same order in your results as you did for the predictions in your introduction and used in the data analysis section of the methods.

Write as clearly as possible. Avoid general statements about groups being *significantly different*. Instead quantify the difference and state the direction (e.g. *males were 10% larger than females, a statistically significant difference (Table x)*).

Support your statements with data. Present data in tables or figures where appropriate. Report summary rather than raw data. Be consistent and sensible with the number of decimal places you use (Section 6.15).

Honest reporting includes reporting the full outcomes of all statistical analyses, including alternative tests of the same hypothesis and all the variables you tested. In other words, don't pick the results that best support your argument, or which are most surprising (Section 3.5).

When presenting the results of a statistical test, include the name of the statistical test, the test statistic and its value, the degrees of freedom (df), or sample size (n), depending on which is most appropriate for that test and the exact p value (not merely whether it is below a threshold or not), unless it is very small, in which case use $p < 0.001$. Present the results of all tests, including those with non-significant outcomes. For example, (Wilcoxon signed-ranks test: $n = 20$, $Z = 3.82$, $p < 0.001$; ANOVA: $F = 2.26$, df = 1, $p = 0.172$). There are various formats for reporting this information (e.g. df are often included as subscripts to the test statistic). Choose a format and be consistent.

If you use arbitrary cut-offs for significance, use them accurately. Your results are either significant ($p < \alpha$) or not ($p \geq \alpha$). Do not round a non-significant result (e.g. $p = 0.052$) down to $p = 0.05$, and then claim it is statistically significant. Resist the temptation to use one of the hundreds of ways to suggest that your findings are *almost* significant.[2]

---

[2] https://mchankins.wordpress.com/2013/04/21/still-not-significant-2/ [Accessed 9 January 2019]. My favourite is *flirting with conventional levels of significance (p > 0.1)*.

Remember absence of evidence is not evidence of absence if a test is non-significant (Section 5.5). Write $p < 0.001$ not $p = 0.000$.

Remember that a p value does not measure the magnitude of an effect and include information concerning the real-world, as well as statistical, significance of any findings by presenting effect sizes (Chapter 5). Report the size and direction of estimated effects even if they are statistically non-significant. Provide confidence intervals to quantify the precision of estimates (Section 5.8). Report the power of tests, calculated *a priori*, not from observed effect sizes or p value (Chapter 16).

If you use model building, model selection, or multi-model inference (Section 15.8.3), report the results of all models.

22.6 PRESENTING YOUR DATA

Think carefully about the most effective way to present your findings to your readers. Never just copy and paste the output from a statistics package into your report. All figures and tables should serve a purpose. They should be clear and easy to understand.

Refer to all tables and figures in the text and explain the patterns you want the readers to observe (e.g. *males were three times the mass of females (Table x)*). Don't repeat the caption in the text (e.g. *Table x gives the results of a statistical test* or *Results are summarised in Table y*). Don't repeat information you provide in a table or figure in the text. For example, there's no need to report the full results of a statistical test in the text if they are also in a table, and there's no need to give summary data in the text if they are also in a figure. Just cite the table or figure so the reader knows where to find the information.

Number figures and tables consecutively and separately.

Give each table and figure a concise caption describing what it shows. Define all abbreviations and terms in the caption, or in explanatory notes below a table, using the same terminology as used in the text. If your findings are for a particular species, study site, and time period, include this information in the caption. Some captions also interpret or comment on the contents.

Tables and figures should be self-explanatory. Your readers should be able to interpret tables and figures without referring to the text (or worse, your brain) for information because many readers jump straight to these elements of an article. For example, *Influence of the treatment on the subjects* forces readers to read the text to find out what the treatment was and which subjects you studied. The definite article (*the*) assumes that a noun is familiar to readers, so don't use it in captions.

Some journals allow or require you to include figures and tables in the text for peer review, although you will probably need to separate the figures and tables from the main text for the final submission. If you include figures and tables in the text, place them following the paragraph in which you refer to them, not before.

Fit figures and tables to the portrait layout, if possible, so your readers don't have to turn the page around or turn their heads to view the screen.

22.7 TABLES: SUMMARISING YOUR DATA

Tables are appropriate when you need to present precise values, for lists, and for the detailed results of statistical results. Use them to avoid repetition in the text. Don't use them to illustrate relationships; figures are better for that.

Consider the design of your tables carefully and make them logical and simple. We have all puzzled over tables in journal articles; don't be that author. For example, it is easier to compare columns than rows. Don't split tables into separate sections (e.g. Table 1a and Table 1b). Either make separate tables (Table 1, Table 2) or combine the data in the same columns or rows. Don't repeat information unnecessarily; the point of a table is to avoid repetition. For example, give units in the column headings, don't repeat them with each value.

Make tables in a word processor or a spreadsheet so that you can edit them easily. Use a legible font size. Ask yourself what information is necessary for your readers to understand your findings and cut everything else. Tables do not have vertical lines and have as few horizontal lines as possible. (Look at tables in published articles to see how this works).

Summarise the content of a table in the heading, but don't list the column headings.

Use superscript letters, numbers or symbols to refer to elements of the table in the notes. Don't use abbreviations unless they are necessary, to save the readers looking them up.

Include exact p values, not a summary.

22.8 FIGURES: ILLUSTRATING YOUR FINDINGS

Figures include diagrams, maps, plots, photographs and drawings. They are very important in allowing readers to understand your study and assess your data.

Use a diagram or flow chart to illustrate a process. Ensure that maps include a scale and a compass direction, and that you explain all elements of the map in a legend or the caption.

Illustrate your major points with plots, which are usually easier for readers to interpret than tables or text are. You can provide exact values in an appendix, if needed.

There are many options for drawing figures. R is by far the most flexible. Eliminate all unnecessary, distracting, and misleading visual elements (**chartjunk**) including unnecessary 3D elements, gridlines, and rectangular frames around plots. Check the file format you need early on, to avoid having to replot your data.

Use **bar charts** to compare quantities, not pie charts, because readers can compare heights more effectively than they can angles. Use a line graph to illustrate relationships between continuous variables (Section 5.2). Always show the data as well as the fitted line, so that your reader can examine the distribution. Where sample sizes are small, use **scatterplots** and indicate paired or matched data by joining them with lines. When summarising larger datasets, plot a measure of the central tendency (usually the mean or the median, Box 5.1) and use error bars to indicate the spread of the data (confidence intervals are best). Also include the sample size. (This can be in the caption.)

Box-plots (box and whisker diagrams) showing the minimum, first quartile, median, third quartile, maximum, and outliers are more informative than bar charts. Report medians, not means (Box 5.1), when using non-parametric statistical tests (Section 15.2.1). When reporting non-parametric statistics for paired or matched data (Section 12.4.2) report the median difference, not the median values for each condition. The median difference is not the same as the difference between the medians for each condition.

Good plots illustrate your study design and are intuitively related to your statistical analyses. Data for between-group comparisons (where subjects are in different groups, measured once each, and you are interested in the difference between groups, Section 12.4.1) are relatively easy to summarise as means with confidence intervals. Plotting data for within-subject comparisons (where you measure the same subject under different conditions, and you are interested in differences within the subject, Section 12.4.2) requires more thought. Plotting the confidence interval doesn't reflect the question you ask in statistical tests, because the confidence interval doesn't account for the matched design and includes variation between subjects that is not relevant to your study design. The easiest way around this is to plot the individual data points. If you have too many data points to do this, consider plotting the difference between the means, with a confidence interval, or calculate a condition-specific confidence interval that reflects your statistical analysis. Figures to show data in mixed designs, which include between-group and within-subject comparisons (Section 12.6) require a lot of thought.

Label all axes and include units in the label (use / to indicate the units on an axis because dividing the value by its units gives the dimensionless number that you plotted). Don't repeat units in the axes label and the tick labels. Use an appropriate number of decimal places in tick labels and ensure all numbers along an axis have the same number of significant figures (e.g. 1.0, 1.5, 2.0 not 1, 1.5, 2). Check that the scaling is appropriate, use legible font sizes, and avoid obscure abbreviations.

Maintain the same symbols for the same subject or treatment throughout your article, to help your readers. Use legible font sizes and be consistent across figures. Check colours work in greyscale for printing and copying. If you have two plots in the same figure with the same axes, you don't need to repeat the axis titles. If you intend your reader to compare two plots, place them near one another and use the same size and scale.

If your figure compares groups and illustrates the results of a statistical test, use * to indicate $p < 0.05$, ** for $p < 0.01$ and *** for $p < 0.001$, but ensure that readers also have access to the exact p values. If you show more than two groups, use brackets to indicate that and link the groups you compare or letters to indicate groups that do not differ. Explain clearly how to interpret these elements of your figure in the figure captions. If you plot a regression line, include the equation in the figure or in the caption.

If your axes do not start at zero, either break the axis or state this in the caption. Don't extrapolate fitted lines beyond the data.

Ensure that your figures are honest. If you choose the best example from your data, say so. Otherwise, choose a representative example. Don't modify images, alter lines, or remove data points to change or improve your results; this is fabrication and is research misconduct (Section 3.1).

Describe all elements of the plot in the legend or caption. A common error is to not explain the elements of a box-plot.

If you use a figure that you have copied from somewhere else, include the source and ensure that you have permission to include it.

Critically examine figures in the articles you read to find good and bad examples. Practice spotting chartjunk and unexplained elements of a figure.

22.9 THE DISCUSSION: WHAT YOUR FINDINGS MEAN

Your discussion summarises and interprets your main findings. It assesses whether the findings support or refute the hypotheses you presented in your introduction, examines alternative explanations, acknowledges limitations, considers puzzling results, compares your findings with those of previous studies, and explains what the study adds to our understanding of the research problem. In other words, now you get to say what you think your findings mean and why they are interesting.

Begin with a succinct summary of your major findings. There is no need to restate your aim, unless the journal's Instructions for Authors require this, or you have a particularly long thesis chapter (in which case, ask yourself why it is so long; always consider your readers). Next, include brief discussion of any limitations to your study that influence the interpretation of your findings. Then, consider whether your original hypotheses are plausible or whether you need to refine them. Discuss the most important findings of your study, not every single result; otherwise, your readers will have difficulty in distinguishing the most important findings from the rest.

Address each major finding in a separate paragraph. Use the topic sentence to describe the pattern you found. Then explain how your results relate to the predictions and to the literature that you reviewed in your Introduction. Compare your findings with those of other studies and propose plausible explanations for general patterns or conflicting findings. Suggest additional research that might resolve any contradictions. Then, address the broader implications of your results. Suggest theoretical implications, and any practical implications (if appropriate). Extend your discussion beyond your specific study species to the broader picture. Avoid too much speculation; your conclusions should arise from your results.

What we already knew should appear in the introduction, not the discussion, unless it is a specific comparison with your results, or part of your interpretation. Your discussion should not repeat the results, but it may summarise them. Citing a figure or a table is a sign that you are repeating your results. Your discussion should not include findings that are not reported in the results section.

Remember that your work supports, or does not support, previous work, not the other way around. In other words, earlier work takes precedence over yours. As with the introduction, be polite about other studies. Be clear whether you are referring to your own findings or to those of other studies. Make your reasoning explicit; don't leave it to the readers to understand why you mention something.

As a rule, a paragraph that does not refer to your results does not belong in your discussion. Don't get side-tracked into what you would like to have done, or now know you should have done with the benefit of hindsight, or lengthy speculation that is not supported by your results.

Your discussion may contain conservation or management implications if these arise from your study. Make these a clear subsection so that users can locate them easily. More general comments on the need to conserve your study species are not discussion or conclusions if they are not based on your findings.

End with the broader implications of your results beyond your case study. These conclusions must arise from your findings and must be supported by the data presented. Don't simply offer your opinion of the general area of enquiry. Conclusions can easily repeat the discussion. Limit them to a brief statement of what we knew before and what the study adds to that knowledge. Don't end with a vague statement about the need for more work, because we always need more work.

22.10 THE ABSTRACT: A CONCISE, STAND-ALONE SUMMARY

An article's abstract and title are freely available to all online. Prospective readers will use them to decide whether to obtain and read your article. This makes the abstract one of the most important components of an article. Write it last and edit it carefully.

Journals differ in the detailed instructions for the abstract. In general, an abstract should summarise the entire report, taking the same hourglass shape as the report. Begin broad, with the general research context (authors often neglect this), then narrow to your aim, a concise account of the methods, a clear description of the most important results, and a summary of the discussion. End with conclusions that highlight the broader significance of the report to the field you introduce in the opening sentence.

Make every sentence in the abstract informative. Avoid vague statements. For example, provide an informative summary of the discussion instead of *We discuss the implications of our findings*.

Journals often set a maximum word count for abstracts (usually 250–300 words), which are usually a single paragraph. Some journals split the abstract into sections using headings. The abstract should be complete, without reference to the text. Be brief and specific, use complete sentences, and a logical flow of ideas. Remove all words and phrases that you can cut without altering the meaning. Don't include unexplained abbreviations or terms. Avoid unfamiliar terms, acronyms, and abbreviations. Abstracts don't usually include citations, but if you must include one, then include it in full, because readers may not have access to the reference list. There is no need to include the results of statistical tests, and certainly no need for p values with no other information. Abstracts rarely include the effect size, but they should. Never include information in the abstract that is not in the paper, and don't make claims that are not supported by your findings.

22.11 THE KEYWORDS: HELP READERS FIND YOUR ARTICLE

Most journals ask authors to provide four to six keywords for their article. Search engines use keywords, so select them carefully, so readers can find your article. Short phrases are often more appropriate than single words. Ask yourself what you would look up to find your paper. Are there any alternative terms that you should include? Explore search engine optimisation strategies. For example, keywords traditionally supplement, but do not repeat the title, and this is reflected in journal Instructions for Authors. However, automated searches make it preferable to repeat terms in the title, keywords, and abstract.

22.12 CITATIONS AND THE REFERENCE LIST

Provide citations for all assertions not supported by the data you report (Section 6.16). Cite reviews where appropriate, rather than long lists of articles.

The reference list contains all the literature you cite in the report, and only that literature.

The exact format of citations and the reference list varies with the journal, so check the instructions and recently published articles. If you have no specific instructions, choose a style and be consistent. Include the digital object identifier (**DOI**, a persistent, unique identifier for a document) for each article. You can find this online.

Don't use a personal communication or personal observation (Section 6.16) to support a central finding or claim, because the readers cannot verify the finding or claim independently. Some journals require written proof of personal communications.

22.13 THE ACKNOWLEDGEMENTS

The acknowledgements (with or without the second *e*) are your opportunity to recognise the help you received during your project. Include help that does not meet the criteria for co-authorship (Box 3.2) in the acknowledgements.

Be generous and honest in your acknowledgements. Thank the people who helped you with ideas, information, logistics, data collection, advice, and editing, if they are not co-authors. Check that people you acknowledge are happy to be named. For example, someone thinks who that you ignored his or her advice may not wish to be associated with your report. Stick to concise facts in a journal article. Thesis acknowledgements can be more effusive. Don't *wish to thank* someone; just thank them. Acknowledge comments from reviewers and editors in revised manuscripts. This includes comments on previous drafts submitted to other journals.

The acknowledgements should also include a statement of grant and other support, with the full names of funding organisations. Some funders ask you to include grant numbers. Also include any conflicts of interest (Section 3.10), such as financial or personal relationships that may be viewed as potential sources of bias, unless the Instructions for Authors ask for these separately.

22.14 APPENDICES

You can include information that supports, but is not essential to, your report in **appendices**. Readers should not need to see appendices to understand the article. Appendices appear after the reference list. In journal articles, this information is often published online only, as electronic supplementary material, with a link or reference in the main text. Appendices can include datasets and code (Box 22.4), and multimedia files that illustrate your article. Some short-format journals report methods in appendices.

Label appendices separately with letters (Appendix A, Appendix B etc.). Journals provide instructions for how to label supplementary material. Refer to all appendices in your report.

22.15 CHAPTER SUMMARY

In this chapter, we've seen that:

- A research project is not finished until we have written it up.
- Readers use the title, abstract and keywords to find a report and then decide whether to read it.
- A report centres on the aim.
- The introduction sets the aim in the context of a broad research question, then the specific case study, and ends with the study set-up.
- The methods section explains what we did to address the aim.
- We should report the rationale for selecting the sample size and the power of tests (calculated *a priori*, not based on the results).
- The results section reports what we found.
- We must report findings honestly and transparently.
- We must identify *post hoc* analyses clearly.
- We must report the size and direction of effects with confidence intervals as well as statistical significance.
- Sharing data and code benefits both science and the researcher.
- We use figures to illustrate key findings and tables to summarise data and avoid repetitive text.
- We should communicate our findings clearly and accurately, design tables carefully, and avoid chartjunk.
- The discussion section interprets the findings and compares them with existing knowledge to synthesise what we now know.

Box 22.5 addresses common problems with reports and how to resolve them.

Now you have a report, you can submit it to a journal for peer review. In the next chapter we look at how to do that.

Box 22.5: Common problems in scientific reports and how to resolve them

Section	Problem	Solution
Introduction	Begins with the study species rather than a theoretical question	Begin with the general theoretical concept and research question, not the particular species you chose to study it in
	Begins with a method rather than a theoretical question	Begin with the general theoretical concept and research question, not the method you used to study it
	Each paragraph summarises a different study	Synthesise the results of studies and how they contribute to our understanding of a question; don't simply report what the authors did and found
	Reviews irrelevant material	Report only information relevant to your specific aims
	Lists studies but gives no details of what they found	Synthesise the results of studies and how they contribute to our understanding of a question
	Muddles the general introduction with the specific introduction	Review all the material that applies generally in the general introduction before moving on to your case study
	Aims are scattered through the introduction, often at the end of each paragraph or subsection	Place the aims together at the end of the introduction. Don't scatter them through the introduction

Box 22.5: (cont.)

	The aims appear at the end of the introduction but include further general introduction	Cover the general context to all the aims in the general introduction
	Introduces the original aims, not the study as it was actually conducted	Introduce the study you report, not the study you would have liked to have conducted
Methods	Insufficient details of the sample size and sampling strategy	Provide details of the study subjects and how you selected them, the study dates, and full sampling strategy, including the sample size
	Insufficient details of the statistical analysis	Provide details of how you tested each prediction, in the same order as the predictions, including all the tests you conducted, and how you checked model assumptions
	Inappropriate analyses, questionable research practices, multiple comparisons, or pseudo-replication	See Section 3.5 on problems in reporting statistical tests; Section 15.4 on choosing the right test; Section 21.2 on sticking to an analysis plan
	Confuses a low statistical threshold (α) with rigour, without considering statistical power (β)	Take both the risk of a false positive (α) and the risk of a false negative (β) into account (Section 5.6)
Results	Results section is long, and readers gets lost	Only report results that are directly relevant to the aims. Guide the readers using informative subheadings. Use the order of aims and predictions that you set up at the end of the introduction

Box 22.5: (cont.)

	Exhausts and distracts the readers with the results of null hypothesis statistical testing	Only conduct and report null hypothesis statistical tests that are directly linked to the predictions
	Doesn't describe the basic patterns observed in the data	Describe the patterns observed before diving into the results of null hypothesis statistical tests
	Includes further methods, or repeats the methods	Report what you did in the methods; report only what you found in the results. The exception is if you ran *post hoc* analyses, in which case specify this clearly
	Reports only significant results	Include full results of all statistical tests, not only those that are significant
	Suggests that non-significant findings are *almost* significant	Report results as significant or not significant
	Reports significance without effect sizes	Include the effect size (Section 5.7) and confidence intervals (Section 5.8) as well as the statistical significance of any findings
Discussion	Concentrates on the limitations and failures	Review the limitations briefly. Concentrate on what you found, not what you couldn't do. Cut aims that the study simply could not address
	Refers only to the study species	Consult the broader literature to compare your findings with what we know about the question in other species

Box 22.5: (cont.)

Treats each finding in detail so the readers can't distinguish what is important from what is trivial	Highlight and interpret the important findings, in the context of your aims
Assumes that the hypotheses are correct, and any lack of support is due to limitations of the study, without considering the possibility that the hypotheses may be false	Include the possibility that the hypothesis may be incorrect or require refinement
Suggests that non-significant findings are *almost* significant, or discusses them as though they were significant	If you report results as significant or not significant, discuss them appropriately, and remember to include the effect size
Interprets non-significant results as a lack of a relationship	Interpret non-significant results as an ambiguous relationship
Interprets the p value as a measure of the importance of an effect	Understand that statistical significance is not the same as real-world significance, and discuss both
Repeats the results (often more succinctly than the actual results). A good clue to this is if you refer to your tables or figures in your discussion	Report what you found in the results; report what it means in the discussion
Is unrelated to the findings	Check that each paragraph mentions your study
Extrapolates beyond the limits of the data	Clarify where you extrapolate beyond the population your sample represents and limit speculation

Box 22.5: (cont.)

	Too speculative	Some speculation is okay, but most of the discussion should interpret your findings
General	Sections repeat one another	Report what you did in the methods, what you found in the results, and what it means in the discussion
	Unnecessary statements of the content of a section. For example: Methods: *In this section, I will explain the methods I used . . ..* Discussion: *I will first discuss . . . then discuss . . .*	Rely on the headings of each section; there's no need to explain that the section does what it is intended to do
	Writing is confused and unclear	Follow the guidance in Chapter 6

22.16 FURTHER READING

Clymo, RS. 2014. *Notes on Reporting Research: A Biologist's Guide to Articles, Talks, and Posters.* Cambridge: Cambridge University Press. Includes a lot of useful advice on communicating clearly with your audience, with a rather old-fashioned, historical approach.

Field A, Miles J, Field Z. 2012. *Discovering Statistics Using R.* London: SAGE Publications Ltd. Chapter 4 gives good guidance on plotting data in R.

Hailman JP, Strier KB. 2006. *Planning, Proposing, and Presenting Science Effectively.* Cambridge: Cambridge University Press. Chapter 3 covers writing a report.

Karban R, Huntzonger M, Pearse IS. 2014. *How to Do Ecology: A Concise Handbook.* 2nd edn. Princeton, NJ: Princeton University. Chapter 8 includes writing journal articles.

Kilkenny C, Browne WJ, Cuthill IC, Emerson M, Altman DG. 2010. Improving bioscience research reporting: the ARRIVE guidelines for reporting animal research. *PLOS Biology* 8: e1000412. https://doi.org/10.1371/journal.pbio .1000412. Describes the need to report comprehensive methods for research using laboratory animals.

Loftus GR, Masson MEJ. 1994. Using confidence-intervals in within- subject designs. *Psychonomic Bulletin & Review* 1: 476–490. https://doi.org/10.3758/ bf03210951. Describes how to plot data to appropriately illustrate study design, including an explanation of the logic behind confidence intervals for within-subject designs.

Lovejoy CO. 1981. The origin of man. *Science* 211: 341–350. https://doi.org/10.1126/ science.211.4480.341. An example of the use of personal communication rather than published empirical findings that got past the editor of *Science*: in note 79, the author cites *DC Johanson, personal communication* to support the assertion that *human females are continually sexually receptive*.

Matthews JR, Matthews RW. 2014. *Successful Scientific Writing: A Step-by-Step Guide for the Biological and Medical Sciences*. 4th edn. Cambridge: Cambridge University Press. A detailed guide to writing, with exercises to improve your practice. Enlivened with quotations and cartoons.

Mensh B, Kording K. 2017. Ten simple rules for structuring papers. *PLOS Computational Biology* 13: e1005619. https://doi.org/10.1371/journal.pcbi.1005619. Advice on how to structure a manuscript, including sentences, paragraphs, sections, and the entire document.

Parker TH, Bowman SD, Nakagawa S, Gurevitch J, Mellor DT, Rosenblatt RP, DeHaven A C. 2018. *Tools for Transparency in Ecology and Evolution*. http://doi .org/10.17605/OSF.IO/G65CB. Includes a checklist of questions to help authors maximise transparency.

Weissgerber TL, Milic NM, Winham SJ, Garovic VD. 2015. Beyond bar and line graphs: Time for a new data presentation paradigm. *PLOS Biology* 13: e1002128. https://doi.org/10.1371/journal.pbio.1002128. Reviews problems with data presentation and calls for more complete presentation of data.

Submitting to a Peer-Reviewed Journal

Disseminating our findings tells other people what we found. Publication is a crucial part of the scientific process and avoids duplication of effort. We should publish our findings regardless of whether they support our hypotheses, to avoid publication bias, which arises when researchers don't publish findings because they are non-significant (Section 3.4). We may need to publish to advance our career, but this is not the purpose of scientific articles.

This chapter goes through the process of submitting a manuscript to a peer-reviewed journal (Figure 23.1). Peer review involves the scrutiny and evaluation of our work by experts (Box 3.1). I begin with how to choose a journal and things to consider before you submit. I revisit the need for research integrity and define publication misconduct. Then I explain additional elements you may need to prepare to accompany your manuscript, the cover letter, submission, and the review process. I explain the editor's decision, what to do if your manuscript is rejected, revising your manuscript, and resubmitting it. Finally, I cover what happens after your manuscript is accepted.

23.1 CHOOSING A JOURNAL

Choosing a journal can seem daunting. There are many possibilities, varying in topic, format, readership, acceptance rate, page charges, and perceived prestige.

The key question is who will be most interested in your findings. Are they appropriate for a broad scientific audience or for specialists in your field? As we saw in Section 10.1, some journals report discoveries in any aspect of science; others cover a more specialised subject area (e.g. behavioural ecology, conservation, cognition, ecology, welfare) or

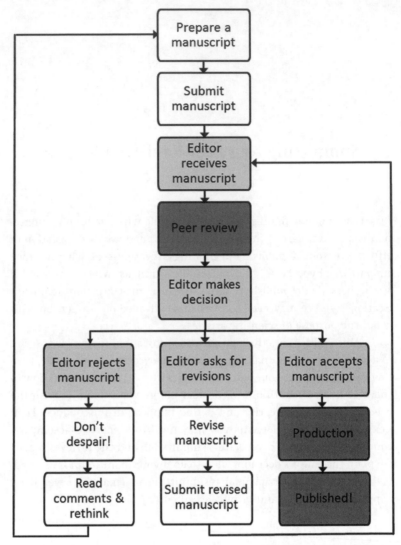

Figure 23.1 The process of submitting a manuscript to a peer-reviewed journal. White boxes are author actions, pale grey boxes are editor actions, dark grey boxes are other actions.

taxon (e.g. primates), with a narrower readership. Choice of journal is important, because you can spend months submitting your manuscript to a journal, receiving a negative response, often without constructive feedback, re-drafting the manuscript for a different journal, and trying again, sometimes several times.

Look at the journals you read regularly, and the studies you cite in your manuscript. Compare the breadth of the question you address and the quality of your data with published articles. If they are of similar quality, then the journal might be appropriate for your manuscript. If in doubt, ask experienced colleagues for help.

Once you have identified some possible journals, check the aims and scope on their websites. Is your topic within the journal's scope? Also check recent tables of contents. Remember that you want people to read your article, so don't bury it somewhere that your audience won't find it.

Does the journal often publish on the topic you work on? If it doesn't but your topic is within the aims and scope, check with the editor to ask whether he or she will consider the manuscript. Don't listen to colleagues who say that a journal is biased against a topic. This information may be out-of-date, and authors may mischaracterise the reasons for manuscript rejection. If in doubt, check with the editor.

Journals are commonly ranked by their impact factor, but these can be highly skewed if some articles are highly cited (Box 10.2). Never use impact factors to compare across fields, evaluate individual articles, or assess the author of an article.

Journals cover publishing costs by charging either readers for access to articles or authors for publishing (Box 23.1). If you find a journal that you have not heard of before and that charges publication fees, or receive an email soliciting a submission, check that the journal is a legitimate publishing entity. Some journals charge publication fees without providing high-quality peer review and editing (Box 23.2).

Further criteria to consider include word limits, whether a journal is owned by a learned society or is purely commercial, the time from submission to publication (this should be available on the journal website), and whether the acknowledgements sections of articles in a journal frequently mention reviewers (a good sign of constructive reviewing).

23.2 BEFORE YOU SUBMIT

Publication is the culmination of a long process, to which many people may have contributed, so most publications have more than one author. Early in your collaboration, negotiate and agree on who will be an author on the manuscript and the order of authors, updating your agreement if needed as the project progresses (Section 2.9). If your target journal does not require a statement of the contributions made to the

Box 23.1: Costs of publishing

Traditional journals cover the costs of publishing through subscription fees or by charging for access to individual articles. Most do not charge authors for publication, although they may charge for colour in the print version of an article. You can work around this by referring readers to the online version but remember that many readers will still print or copy your article in greyscale.

In contrast to traditional journals, open access journals (Section 3.5) charge authors **article-processing fees** (or **publication fees**), payable on acceptance, and make articles freely and permanently available online for anyone to access. Authors retain the copyright. Open access publishing increases the size of the potential audience and improves transparency, and some funders require it. Article-processing fees can be high but some publishers waive the fees for authors from low-income countries or in cases of hardship. Some pay-to-read journals also offer an open access option (**hybrid journals**).

In addition to publishing in an open access journal (**gold open access**), you can make a version of your manuscript freely available to everyone in a disciplinary or institutional repository (**green open access** or **self-archiving**). The publisher usually retains the copyright for these articles and imposes terms and conditions about which version of the manuscript you can deposit, and when it can be made available.

study by each of the listed authors (Box 3.2), you can include this in the acknowledgements.

Obtain consent to submit from all co-authors and any responsible authorities at the institution(s) where the work was conducted before you submit an article. Don't leave this until the last minute; remember that they will need time to read and comment on the manuscript.

Read and comply with the Instructions for Authors for your target journal.

Get feedback on your draft (Section 6.3). Contact experts and ask for their opinion, if necessary (Box 13.2). Act on the feedback you receive; don't simply rebut it by explaining why the reader is wrong. If one reader has a question, or misunderstands, other readers may do so, too. Recognise this feedback in the acknowledgements. Preprint

Box 23.2: Pseudo-journals

The rise of open access publishing (Box 23.1) has been accompanied by an increased number of publications that do not provide the services expected of reputable journals (**predatory journals** or **pseudo-journals**). Such journals may have little or no peer review. They often solicit submissions via email, promising unfeasibly rapid publication, and may have titles, aims, and scope that make little sense.

You can identify pseudo-journals through careful investigation. For example, are you already familiar with the journal? Have you read articles published in it? Do you know anyone who has submitted work to it? Has anyone recommended it to you? Is it indexed in the databases that you use to find articles? Do you recognise the names of the editorial board and recently published authors? Do the members of the editorial board mention the journal on their personal websites? Does the name of the journal make sense, or is it suspiciously similar to a well-known, established journal?

If you receive an email soliciting a manuscript, check that the message addresses you appropriately and correctly (e.g. I receive emails from pseudo-journals addressed to Dr Mandrills).

Also check the journal website for:

- A clear statement of who publishes the journal and their contact details, including where the offices are
- A clear editorial process and peer-review policy
- Membership of the Committee on Publication Ethics (COPE) and other industry initiatives
- A clear statement of any publication charges
- Accurate spelling and grammar

If in doubt, ask a senior colleague for advice.

FURTHER READING

Laine C, Winker MA. 2017. Identifying predatory or pseudo-journals. *Biochemia Medica* 27: 285–291. https://doi.org/10.11613/BM.2017.031. Reviews efforts to assist authors and readers in identifying pseudo-journals and legitimate open access journals and provides a framework for doing so.

Shamseer L, Moher D, Maduekwe O, Turner L, Barbour V, Burch R, Clark J, Galipeau J, Roberts J, Shea BJ. 2017. Potential predatory and legitimate

> **Box 23.2:** (cont.)
>
> biomedical journals: Can you tell the difference? A cross-sectional comparison. *BMC Medicine* 15: 28. http://dx.doi.org/10.1186/s12916–017–0785-9. Identifies evidence-based characteristics of predatory biomedical journals, which are also relevant to other fields, or easily translated.
>
> The "Think, Check, Submit" campaign helps researchers to identify trusted journals. http://thinkchecksubmit.org. Bear in mind that it is run by a coalition of publishers, who may have a conflict of interest.

servers (e.g. bioRχiv and *PeerJ PrePrints*) allow you to make manuscripts available before peer review and get feedback. They are little used in primatology, but allow you to establish priority for your work, as well as soliciting feedback, and their use is likely to increase.

23.3 PUBLICATION ETHICS AND MISCONDUCT

When preparing a manuscript, reread Chapter 3. Remind yourself of the definitions of fabrication, falsification, and plagiarism, and avoid research misconduct. Avoid questionable research practices by acknowledging and crediting relevant work, reporting methods honestly and fully including details of all data manipulation and analysis, avoiding selective reporting, assigning authorship credit appropriately and disclosing any conflicts of interest.

In addition to these responsibilities, avoid the following types of publication misconduct:

Text-recycling or **self-plagiarism** involves reusing sections of your own publication in another publication. Citing the original source does not justify doing this. The exception to this is that we often provide similar details of our study taxon and methods in multiple manuscripts. In such cases, we should clearly identify that earlier work (e.g. *As in previous analyses . . .*).

Redundant or **duplicate publishing** involves including information in a manuscript that you have already published without clear reference to previous work. It is almost never appropriate to include previously published findings in the results section, although you may reuse a dataset if the new work expands on your previous work and the previous work is cited fully and clearly. The citation must include an explanation of which data have been used in previous analyses. The new manuscript must include sufficient new material to justify publication. In other words, minor reanalysis to address a very similar question is not appropriate. Some journals will not allow reuse of a dataset, even if you use it to address a new question.

We often publish more than one article from a single study. However, when the findings can be presented together in a single manuscript of regular length, then we should combine them. Splitting a single study into several parts to increase the quantity of manuscripts we publish is termed **salami-publishing** or a **least publishable unit** strategy.

23.4 HIGHLIGHTS, GRAPHICAL ABSTRACTS, AND LAY SUMMARIES

Many journals either require or give the option of additional components to help reach a broader audience. You may need to submit these with your manuscript, or you may be asked for them once your manuscript is accepted. Preparing these is very useful in identifying your key points and may lead you to revise your manuscript but remember that you must be open and transparent in reporting science; tell the full story, not a selective account.

Graphical abstracts are concise, visual summaries of the main findings in a single panel. They are designed to draw attention to the article when potential readers search online or browse online tables of contents and are often used on social media. A good graphical abstract conveys an immediate understanding of your take-home message to the reader. Designing this visual summary requires careful thought. Follow the publisher's guidelines and use examples for inspiration. Tell stories from left to right or from top to bottom. Include only key information, be sparing with text, and avoid unnecessary clutter. Get feedback to clarify your message.

Highlights are short lists of bullet points (often 3–5 points) that communicate the core findings of an article. They appear in the online Table of Contents, and at the beginning of the article. Check the guidelines for whether the highlights should be written for experts in the field, scientists more generally, or the general public. Be clear and concise and use simple language. Reflect your findings accurately.

Lay summaries are plain language summaries of an article for the general public. Explain the problem you addressed, the key results, and what they mean. Avoid technical terms and disciplinary jargon. Don't overstate your findings. Ask someone who is not expert in your field to check that your lay summary makes sense to them.

23.5 THE COVER LETTER AND SUGGESTING REVIEWERS

You may need to include a **cover letter** when you submit your article to a journal. Journals may give instructions on the content of the letter if they require one. For example, you may need to include confirmation that the manuscript has not been published elsewhere and is not under consideration by another journal, and that all authors approve the submission.

The cover letter is an opportunity to explain to the editor why you think he or she should consider your manuscript for publication. It is formal letter, on headed notepaper, with the date and the address of the recipient. Address your letter to the editor. Look up his or her name on the journal website, spell it correctly, and get the title right. If the title is unclear, use Dr, which has the advantage of being gender-neutral.

The body of the letter should be four short paragraphs. In the first paragraph, introduce the manuscript you are submitting, including the title and the authors. In the second paragraph, briefly describe your aim and your major findings. In the third paragraph, explain why you chose the journal. Refer to the aims and scope and explain why you think the journal's readers will be interested in your work. You can also list a few recent studies published in the journal that are similar to your own. In the fourth paragraph, thank the editor for considering your work.

You can suggest reviewers for your manuscript in your cover letter, or during the electronic submission process. Suggest reviewers who are knowledgeable on the topic and whose advice you would value. Briefly explain why you think the person would be a good reviewer. Don't list your friends or famous primatologists unless they have relevant expertise. Don't suggest people that you have acknowledged – the editor will spot this. You can also name people who should not be asked to comment on your work because they have a conflict of interest. If you do this, give a brief explanation as to why. This is not an opportunity to attempt to block all the experts in your field who might detect flaws in your work.

23.6 SUBMISSION

Read your manuscript carefully before submission. Reread the journal requirements and check that you have followed them. Some journals send submissions back repeatedly until you have followed every point they make, which can be very frustrating. It's easier to simply follow the instructions straight away.

Submit your manuscript.

23.7 THE REVIEW PROCESS

Your manuscript first goes to an editor who checks that it meets basic criteria for the journal. This might be the Editor-in-Chief, or this task may be delegated to an Associate Editor. The exact titles and roles of the editors vary from journal to journal.

If the editor judges a manuscript inappropriate for the journal, he or she may reject it at this stage. Reasons for rejection include manuscripts that fall outside the aims and scope of the journal, findings that do not advance our understanding of a topic, poor study design, poor reasoning, inadequate sampling, inappropriate data analysis, or ethical concerns. If the methods are unclear or lack detail such that readers cannot assess the data, the editor may invite you to revise and resubmit your manuscript, or may reject it.

If your manuscript passes the initial editorial evaluation, the editor will invite expert reviewers to evaluate it, usually aiming to obtain advice from two or three researchers. This can take some time, because reviewers offer their services voluntarily.

The reviewers read the manuscript carefully. The purpose of peer review is to improve the quality of the manuscript under review, and of the material that is eventually published. A good reviewer is constructive and seeks to improve the final report. Reviewers advise the editor. You may hear people say that they reviewed a manuscript and rejected it. They shouldn't tell anyone this, because the review process is confidential, and they didn't reject the manuscript, they made a recommendation to the editor.

Once the reviewers have submitted their reports, the handling editor reads and evaluates both the reviews and the manuscript. The editor then has two options: either request revisions from the authors, via the corresponding author, or make a recommendation of acceptance or rejection to the Editor-in-Chief, who then issues the final decision.

The journal may send you electronic notifications as your manuscript moves through the system and you can often monitor progress in the online system. If the review process is unusually long (i.e. considerably longer than advertised on the journal website), send a polite enquiry to the editor.

23.8 THE EDITOR'S DECISION

A first decision of *accept as is*, with no requests for revision, is very rare indeed. You may receive notification of *acceptance subject to minor revisions*, with a list of concerns to address. **Minor revisions** are usually changes to the text. If you make these, then your manuscript will be accepted. **Major revisions** may include reanalysis of your data, or additional data collection. A request for major revisions does not guarantee acceptance. Instead, the editors will reassess your manuscript following revision, and may seek the advice of reviewers concerning the revisions and response to the comments on the initial submission. Editors may reject the manuscript but invite you to resubmit it if you can address all the concerns raised by the reviewers. In some journals, this **revise and resubmit** option has replaced major revisions.

23.9 WHAT TO DO IF YOUR MANUSCRIPT IS REJECTED

If your manuscript is rejected, you are not alone. It happens to all researchers. Read the comments carefully. It's not usually the end of your manuscript, although sometimes you might need to start again if the reviewers have detected a serious flaw in the study.

If you are sure the editor's decision is wrong, write to him or her to explain why.

Revise your manuscript based on the reviewers' comments and submit it elsewhere. Don't just reformat and resubmit it elsewhere without any revisions. The same reviewers may well review it again. If you have ignored points that are easily addressed, reviewers will not be well-disposed towards your submission.

23.10 REVISING YOUR MANUSCRIPT

You might think your manuscript is perfect when you submit it, but reviewers very rarely agree that it is, so you will probably need to make at least one round of revisions before it's accepted. Don't give up.

The review process is designed to improve the final article. Reviewers usually try to be helpful, but they are busy people. In an ideal world, all reviews would be constructive and helpful. Most are. However, some reviews are abrupt and direct because reviewers are busy, and some are unconstructive (satirised on the Internet as 'Reviewer 2').

When you receive the reviewers' comments on your manuscript, breathe deeply and skim-read them. You may then wish to put them away and read them again more carefully when you have recovered from your initial reaction. At that stage, ask yourself whether the comments are reasonable. Note what you can do to address each point.

Make the suggested changes to your manuscript, using this as an opportunity to improve the final submission. Explain all your revisions in a letter to the editor (**response letter**). Begin the letter by thanking the reviewers, even if you must grit your teeth to do so. Then, copy the comments into your letter, in the same order as they were made. Respond to each point carefully and in full, including line numbers so that the editor can find the revision easily in the manuscript. Simply adding *done* is not helpful unless you have corrected a typo. If you choose not to change your manuscript in response to a comment, explain why diplomatically. For example, you may have a convincing argument for why the revision is not needed. Avoid anger, sarcasm, or counterattack and don't make your response an exercise in rebuttal. It's not in your interests to antagonise the editor. Moreover, editors may pass your revision and comments back to the reviewers for their advice.

If the reviewers misunderstood what you wrote, then the text was not sufficiently clear, and you need to revise it. Don't just claim that the reviewer is stupid.

Don't over-interpret the comments and do things the reviewers and editor didn't ask you to do. If you do spot errors and make changes in addition to those your reviewers require, explain this to the editor in your response letter.

Only the editor, and possibly the reviewers, see your response letter, so if you respond to a reviewer comment in your decision letter, also make the appropriate change to the manuscript, as other readers may well have the same concern as the reviewer.

23.11 RESUBMITTING YOUR MANUSCRIPT

When you resubmit your revised manuscript, the editor will assess it and your response to the reviewers' comments. If the changes are substantial, the editor may invite the reviewers to reassess it or may invite fresh reviewers to examine it. You may receive a further set of comments to address.

23.12 ACCEPTANCE AND PUBLICATION

If you have addressed all the comments on your manuscript, the editor will accept it for publication. Hooray! Celebrate, tell your co-authors, thank other contributors, and notify your funders.

The editor will forward your final manuscript to the publisher or instruct you to do so. Accepted manuscripts may be reviewed for correct style and language (**copy-edited**). They then proceed to a preliminary version of the article (**proofs**). Proofs come with detailed instructions and include questions for the author. Respond to these and check the proofs carefully for errors. The copy editor may have inadvertently changed the sense of what you wrote. Examine individual words; you will naturally skip over them as you read. Check the tables and figures carefully, too. Correct errors, but don't make new revisions. If you can, get a colleague to check the proofs, too.

If possible, make your data and code available in a permanently supported, publicly accessible database when your article is published (Box 22.4).

You may also receive other emails from the publisher, including a copyright release form, an order form for reprints, and an invoice for any publication charges. Journals may also accept suggestions for cover images.

Once your article is published, send a copy to your co-authors, other contributors, and your funders. Use social media to let people know about your article and to promote discussion of your findings.

If you discover an error in your article once it is published, notify the editor, and use a correction notice to correct a small portion of an otherwise reliable publication or a retraction where data are seriously flawed, and the findings and conclusions cannot be relied on (Section 3.6). Corrections and retractions maintain the accuracy of research record.

23.13 CHAPTER SUMMARY

In this chapter, we've seen that:

- Disseminating our findings is part of the scientific process.
- We must make our results available to other researchers to avoid duplication of effort.

- We choose a journal based on the audience we wish to reach and the relevance of our findings.
- We should read and comply with Instructions for Authors when preparing a manuscript for submission to a journal.
- We must avoid research misconduct, questionable research practices, and publication misconduct.
- Manuscripts undergo peer review to evaluate the findings.
- Reviewers should provide constructive advice that will improve the quality of the final report.
- Editors make decisions to reject a manuscript, request revisions, or accept it based on the reviewers' reports and their own reading of a manuscript.
- Editors reject manuscripts that are unsuitable for the journal.
- Almost all manuscripts are sent back for at least one revision before they are accepted.
- Requests for revisions can be painful, but we should engage with comments constructively.

In addition to written reports, we also present results at scientific conferences. This is an excellent opportunity to discuss our findings with other researchers and get inspiration for future work. In the next chapter we look at how to present results at a conference.

23.14 FURTHER READING

bioRxiv (pronounced *bio-archive*): a free online archive and distribution service for unpublished preprints in the life sciences. www.biorxiv.org.

The Committee on Publication Ethics. http://publicationethics.org/. Promotes integrity in research publication. Primarily aimed at editors of scholarly journals, but a very useful source of information on publication ethics for authors.

Hailman JP, Strier KB. 2006. *Planning, Proposing, and Presenting Science Effectively*. Cambridge: Cambridge University Press. Chapter 3 covers submitting a manuscript to a journal.

Matthews JR, Matthews RW. 2014. *Successful Scientific Writing: A Step-by-Step Guide for the Biological and Medical Sciences*. 4th edn. Cambridge: Cambridge University Press. Chapter 17 covers publication.

Roig M. 2015. Avoiding plagiarism, self-plagiarism, and other questionable writing practices: A guide to ethical writing. https://ori.hhs.gov/avoiding-plagiarism-self-plagiarism-and-other-questionable-writing-practices-guide-ethical-writing [Accessed 9 January 2019]. US Department of Health and Human Services Office of Research Integrity educational module to help identify and prevent questionable practices and promote ethical writing.

Instructions for Authors for your target journals.

Publisher's websites for guidance on graphical abstracts, highlights, and lay summaries.

24

Presenting Your Work at a Conference

Conferences are excellent opportunities to hear about the latest news in your field, or in your broader discipline. They are also a great chance to meet like-minded people, share experiences, discuss ideas, and gain inspiration. Friendships, collaborations, new research directions, invitations, and job offers often arise from conversations at academic conferences.

In this chapter, I first introduce academic conferences, and cover preparing and submitting an abstract, then attending a conference. Next, I provide general advice on presentations, then cover preparing and presenting oral and poster presentations. I end with conference etiquette. I focus on presenting at academic conferences, but cover public engagement briefly in Box 24.1.

24.1 CONFERENCES

Conferences range from small regional or national gatherings to huge international meetings. They may address a particular topic or may be a general meeting of a learned society, including symposia on a variety of topics as well as society business meetings. Conferences may have only one session at a time, with all delegates in the same room, or may have multiple concurrent sessions in a conference venue where all the rooms and corridors look the same and it's very easy to get lost. Most conferences include **keynote** or **plenary** presentations by major researchers in the field. This is a great chance to meet the people whose articles you have read and admired.

Box 24.1: Public engagement

Public engagement provides an opportunity to disseminate your research to a broader, non-academic audience. It promotes public awareness of research, gets people excited by science, and can challenge stereotypes of scientists. Engagement can benefit primate conservation by raising awareness of primates and their status. It benefits us, too, by improving our communication skills and helping us understand the broader context of our research.

Opportunities for public engagement include appearances in the media (e.g. on television and radio), public lectures, more informal conversations in Science Cafés, science fairs or festivals, visiting schools, and inviting people to visit your lab or field site. You can make video podcasts about science, get involved in science comedy, or dance your PhD.

Media Appearances
Invitations to appear in the media can come at short notice, because journalists work to tight deadlines. Media training is very useful for interviews, if it is available to you (e.g. through your institution). Interviews may be in person in a studio, on the phone, or in a remote studio when the interviewer is in a different location. Interviews can be live or recorded. Journalists often ask questions that include the answer, particularly if you record a segment several times. Don't let this put you off, and don't simply reply *yes*, because the question may be edited out later, leaving only your answer. Instead, answer in your own words, and include more information or an example.

Remember that you're expert in your subject and be confident. Be clear and concise. Use non-technical language and everyday analogies. Journalists may also change the subject to current affairs or ask about a different study that they're familiar with, so you may need to think on your feet. If you don't know the answer, it's fine to say so. Some reporters like controversy, so consider how you will respond to tough questions.

Public Engagement Events
When preparing for an event, identify your audience. How many people will there be, of what ages, and what level of understanding and knowledge are they likely to have? Do they have any specific

Box 24.1: (cont.)

needs? If you're visiting a school, ask the teacher what students already know.

Choose a catchy, exciting title for your event. You might present your specific project or you might choose a broader topic. Consider what you want your audience to learn and design your presentation to convey that understanding. As with writing for non-specialist audiences, use everyday, non-technical language (Box 6.1). Don't patronise your audience.

As primatologists, we have the great advantage that our study species are highly charismatic. Share your enthusiasm and passion for primates. Use stories and vivid analogies to engage your audience. Provide take-home messages and actions, if appropriate. Be prepared for a broad variety of questions.

Most outreach events reach audiences who are already actively interested in science and seeking information. If you aim to reach other audiences, collaborate with individuals and organisations expert in science communication.

Conservation Education

If your project involves a conservation education programme, and particularly if it is crosscultural, it is essential to work with, and learn from, expert educators and your target audience, to design materials that are effective and appropriate to your audience. If you visit schools work with the teachers. Read the literature on planning, implementing, and evaluating conservation education programmes. Pilot your materials and evaluate their effectiveness. In other words, conduct conservation education in the same way you would any other part of your project. Naïve or poorly informed actions may be ineffective or even detrimental to conservation.

FURTHER READING

http://gonzolabs.org/dance/. Website for the *Dance Your Ph.D.* contest, which challenges scientists to explain their research with dance.
Jacobson SK, McDuff M, Monroe M. 2015. *Conservation Education and Outreach Techniques*. 2nd edn. Oxford: Oxford University Press. A very useful resource for planning, implementing, and evaluating conservation education and outreach programmes.
National Academies of Sciences, Engineering, and Medicine. 2017. *Communicating Science Effectively: A Research Agenda*. Washington, DC: The National

Box 24.1: (cont.)

Academies Press. https://doi.org/10.17226/23674. A report on the science of science communication and the need to go beyond simply conveying information. Freely available online.

Varner J. 2014. Scientific outreach: Toward effective public engagement with biological science. *BioScience* 64: 333–340. https://doi.org/10.1093/biosci/biu021. Describes common misconceptions that can undermine science communication and advocates a scientific approach to science outreach.

24.2 PREPARING AND SUBMITTING A CONFERENCE ABSTRACT

Presenting your own research at a conference raises your personal profile and allows you to obtain feedback to improve your work. Conferences usually allow researchers to submit abstracts for oral and poster submissions. For larger conferences, you can also propose a **symposium** with several speakers on a topic. This is a great way to get people together to discuss a topic and develop your network. Calls for symposia can be made up to a year ahead of the conference, and you will need time to put together a strong proposal, so look at the conference website well in advance. Invite diverse contributors including people of different genders, from various career stages, and from under-represented groups. Actively seek people to invite. Don't just pick the names that come to your mind first because this facilitates unconscious bias (Section 4.11). Not everyone you invite will accept your invitation, so plan for this, too.

Abstract submission deadlines for major conferences may be months before the event, to allow the organisers time to review submissions and plan the conference – a major undertaking. When preparing an abstract, follow the instructions on the conference website. Make your title concise and informative. Provide details of the authors and their affiliations (Box 3.2). Write a summary of your work that can be understood without reference to any other information. Use full sentences and correct spelling and grammar. Include the context of your study, the aim, the research question, methods, results, and discussion. Some conferences may require p values in your abstract. By now, you should understand why these are meaningless without effect sizes, so also provide effect sizes (Section 5.7). If you do not yet have your full results, then either include preliminary results, or give details of the hypotheses and predictions and how you would interpret support for these. Don't use empty phrases like *Results will be discussed in the light of the literature.*

Allow your co-authors sufficient time to provide feedback on your abstract.

Some conferences offer the option of including figures in your abstract or creating a graphical abstract (as for an article, Section 23.4). These can be very effective if carefully designed.

24.3 ATTENDING A CONFERENCE

Many societies provide substantial discounts and bursaries for students attending their conferences. Conferences may be held in expensive city hotels, but most will also provide links to less expensive accommodation, or you can find this online. If you're attending on your own, then you can use social media to link up with other delegates and share accommodation.

If you're presenting, make sure you know when your presentation is and check the instructions for uploading presentation files or putting up your poster. If the meeting is large, with concurrent sessions, review the programme in detail and make yourself a schedule in advance. You can often do this using the conference's online system.

Meeting other researchers is one of the foremost functions of a conference. This can be challenging, particularly for introverts. Conference days are often long so if you know that you need time to recharge, make sure you plan for this (you won't be the only one). Wear your name tag at the conference, attend social functions, and talk to people. Ask questions, compliment their presentation (if relevant) and listen to what they say. Don't just talk about yourself. Never assume you know more than the person you're talking with. If you are nervous or shy, remember that it's very likely that the other person is also ill at ease. If you're attending with your mentor, ask him or her to introduce you to people that you would like to meet. Your mentor may look busy catching up with friends, but introducing you to his or her network is part of the mentor's role. Schedule meetings with people you know you want to talk with; otherwise you might get to the last day and discover the person you were keen to talk to has already left. Take business cards with you and use them to stay in touch with people after the conference. Follow-up on contacts you make after the conference.

Societies often arrange mentoring sessions, workshops, and mixers at conferences. If you find such events difficult, set yourself a target of talking to at least one person. It may look like everyone else there knows everyone else while you don't, but if you look around you will see other people in the same position as you. Talk with one of them. Remember that you have the topic of the conference in common.

Sadly, many conference programmes are male-biased. This is both caused by, and leads to, unconscious bias in favour of men. If you come across an all-male panel, offer the organisers positive and constructive advice on how to avoid a gendered meeting. (There is plenty of material on the Internet.) Conferences also vary greatly in how much support they provide for childcare, although this is improving. If the organisers don't provide information on childcare, and you need it, ask for it. Taking a co-carer (e.g. a partner, parent or friend) to the conference can be less stressful and not much more expensive than childcare. Conferences should also provide lactation rooms for mothers. If they don't and you need one tell the organisers what you need (a private space, comfortable chair, water supply, etc.).

24.4 GENERAL ADVICE ON PRESENTATIONS

Presentation design is a matter of personal style. Nevertheless, some points apply generally. The key is to communicate your aim and findings and why they matter to your audience as clearly and simply as possible. Free yourself from the formal structure of a scientific report (Box 11.1). You don't need details of your methods, particularly if they are conventional. Those who are interested can ask you for the details. You may wish to mix results and discussion to get your message across clearly. Note what you like and don't like about other people's presentations and use this to improve your own.

Consider who your audience will be, how many people are likely to attend, and how much they know about your topic. If your audience is a general one, then you will need to include more background material to orient them, and less technical detail than you would for a specialist audience.

Check the instructions carefully and note the format and scheduling of your presentation. If you use a presentation programme to design your slides or poster, set the slide size before you begin. Changing the proportions later will distort the graphics and text.

Decide on the key message you wish to communicate. Get one idea across well rather than several badly. You want your audience to understand and remember what you say. Tell your audience how your study fits into a broader picture, and why they should be interested.

Only provide relevant information, not everything you know about a topic or a species.

Avoid excess information and distractions in presentations. Use graphics rather than text and tables. Don't just copy figures and tables directly from a paper or from a standard statistics package. Simplify them to show only the essentials. Use colour with caution, and to guide your reader, rather than for decoration. Remember that some people are red-green colour-blind. Give the source for any image you use. If you use images from the Internet, then set your image search to find images that are free to use or share.

Avoid overcrowding a slide or a poster and reduce the text as far as possible. Where you do use text, make it concise and readable. You don't need full sentences. Avoid elaborate fonts. Fonts without serifs (small lines attached to the end of the strokes in letters) are designed for headings (e.g. Arial, Calibri); those with serifs are for text (e.g. Cambria, Times New Roman). Write titles in sentence case, which is easier to read than ALL CAPS or Title Case. Check carefully for typographic errors.

Both oral and poster presentations are valuable opportunities to get feedback on your work via questions from the audience or viewers. Check that you understand a question and ask for clarification if necessary. Answer briefly, directly, and politely. Questioners are rarely rude, but if they are, courtesy is the best response. If you don't know the answer to a question, say so, and suggest how we might find out. If one questioner attempts to dominate the discussion, suggest politely that you discuss the topic in person after the session.

24.5 PREPARING AN ORAL PRESENTATION

An oral presentation at a conference is a great chance to present your work to others working in your field, obtain feedback, and raise your own profile. Most oral presentations use visual aids, usually in the form of a slide show. Start your presentation with a slide that shows your title and name clearly. Include your contact details.

Put your acknowledgements at the beginning of your talk so that you can end with what you want people to remember, rather than a list of names and funders. Alternatively, include photos of your collaborators during the talk, and use logos for funders.

Set up the structure of your talk for your audience, so they know what to expect, but don't simply say that you'll present introduction, methods, results, and discussion. You probably don't need an outline for a short conference presentation, but you might for a longer talk. Repeating your outline slide during your talk helps to guide your audience, or you can repeat the structure along the top or one side of your slide, highlighting the current section.

Present one main idea per slide and summarise it in the title. Build complicated slides up using simple animations. Don't distract your audience with your visual aids. If your audience is reading text on a slide, trying to figure out a complex graphic, can't read a table copied from an article, or is distracted by unnecessary gimmicks, then they will not be listening to you. Place key references at the bottom of the slide.

Don't use complicated tables. Don't be the presenter who puts up a large table of tiny text and admits immediately that he or she knows the audience can't read it.

Check that your slide content is visible, even in poor lighting. Use high contrast and check colours carefully. Black text on white is easiest to read.

The number of slides you need varies with their content and your presentation style, so practice to determine how many you need. People often begin with one slide per minute for a short talk, but fewer for a longer one. If you have additional material that might interest your audience, include it in extra slides after your final slide.

Preparation improves confidence. Rehearse your presentation, get feedback, and refine it. If you tend to feel nervous, on your first slide place an image that helps you to relax and reminds you that you know what you're talking about (e.g. your favourite study animal).

24.6 PRESENTING AN ORAL PRESENTATION

Most conferences provide instructions for when to upload your slides in advance of your talk. Follow these instructions. Familiarise yourself with the room you'll be presenting in, including the slide advancer, microphone, and pointer. Don't assume that you will be able to view the presenter notes in your presentation file; many audiovisual systems don't have this function.

Arrive in plenty of time for the session and introduce yourself to the session chair. Sit near the front of the room. When it's your turn to present, take a deep breath, thank the chair for introducing you and the audience for attending. Speak slowly and face your audience, not the computer or screen. Focus on people in the audience who look interested (but don't stare). Don't be put off if people look uninterested; it probably has nothing to do with your presentation. People who look like they are texting or who are using a tablet are often taking notes.

Use the microphone if there is one. Some people in the audience may have impaired hearing.

Use simple language appropriate for an international audience. You may be tempted to write yourself a script, but don't. We use more formal language when writing than when we talk and reading from a script is far less engaging than a more informal presentation using your slides as prompts. If you must have notes, use cards, and make sure they're clipped together in case you drop them. A printed copy of your slides is a useful prompt in case the projector fails (it happens).

Explain your slides, but don't read out any text on them; your audience can do that for themselves, and they can read faster than you speak. Make sure what you say relates to what you show. Guide your audience through your presentation carefully and explain your figures in detail. For example, walk your audience through each plot slowly, pointing out how to read them, what the axes show, and what you want them to notice.

Use the phrase *In conclusion* ... to tell your audience that you are nearly done and make sure you don't make any further points after the conclusion (another good reason for placing your acknowledgements early in your presentation).

If the remote control doesn't work, check that you're pointing it at the receiver. Don't point laser pointers at your audience.

If your mind goes blank during your presentation, don't panic. Collect your thoughts and breathe. Your audience will appreciate the break.

Allow plenty of time for questions. It can be tempting to run right to the end of your time slot to avoid questions, but you might miss out on very useful feedback. The session chair should take questions for you, but if you do this yourself, ensure you take questions from the whole room and that you don't concentrate on particular groups (e.g. senior researchers or your friends). Repeat the question to the audience if they didn't hear it.

Questions may be points of clarification and details of your methods or analysis, broader theoretical questions about the context of the study, or interesting information or ideas that will enhance your study. To answer questions, it can be useful to go back to slides in your presentation or use the extra slides you prepared. If you need time to think about your answer, reply with *That's a very interesting question, thank you* while you think. If you don't know the answer to a question, that's okay. Stay calm and say that you don't know, or that you need to think about the question. Suggest ways to find out the answer or offer to talk with the questioner after the session. Don't pretend you know the answer to a question when you don't and don't answer with irrelevant information, because this gives a poor impression. Unanswered questions are a source of ideas for future research and collaborations.

Some questions are really statements intended to focus the audience's attention on the questioner. If this happens, thank the person for his or her point and respond if you have something useful to add. Questions are rarely hostile, but this does happen. If it happens to you, stay calm and remember that it's not about you. If you can answer their question, do so calmly and politely. If you can't, then thank the person for the question and move on.

On occasion, questions develop into a conversation, a debate, or even an argument. The chair should step in to avoid this but if he or she doesn't, suggest that you continue the conversation after the session, so that other people can ask their questions.

There may be no questions; that's okay, too. After your talk, remain in the room until the next break. There may be people in the audience who want to talk with you.

24.7 PREPARING A POSTER PRESENTATION

Presenting a poster is an excellent opportunity to discuss ongoing work with interested people, and to exchange contact information for further discussion or collaboration. Poster sessions are often crowded, and may include hundreds of posters, making poster presentations initially more about competitive selling than other presentations, as you try to attract viewers. However, it is your ideas that retain viewers, not your sales skills, and you will benefit more from one or two detailed discussions of your work than you will from crowds of visitors.

Check the instructions for size guidelines before starting to design your poster. Posters are visual, so use graphics wherever possible. Place a clear title at the top of your poster, ideally so that it is visible over the heads of viewers. Decide on the key findings you wish to convey and present these graphically. Don't try to cram in too much information. Make the figures self-explanatory, with all the relevant information, including very brief methods and interpretation. A research article cites figures that are found elsewhere in the article for practical reasons; you don't need to do this with a poster.

Consider the flow of ideas and use arrows or numbers to guide your readers through your poster. We usually read left to right and top to bottom. Use white space carefully for balance, visual relief, harmony, emphasis, and to direct the viewer's eye. Most viewers will not read from beginning to end.

Where you use text, keep it short. You don't need to repeat the abstract on your poster. Make any text easy to read from a distance. Use the active voice, not the passive (**Section 6.8**). Use headlines, and short, concise bullet points, not paragraphs. Use a consistent serif font, line spacing just larger than 1, and left-justified rather than fully justified text, for ease of reading.

Include a prominent **take-home message** (what you want people to remember). Add your contact details. A photo of yourself is useful, so that people can identify you. Include your acknowledgements in a box. Use logos to acknowledge funders. Include key references only. Abbreviated citations in the text mean the reader doesn't have to look elsewhere on the poster for information. Alternatively, list references in small print at the bottom of the poster. Barcodes (e.g. Quick Response codes) are useful to link to further information, supporting videos, copies of the poster, and so on.

Once you have a draft of your poster, check it carefully for unnecessary and redundant information. Put yourself in the place of your viewers. Get feedback and revise your poster for clarity and simplicity.

Finally, allow enough time to print your poster. Fabric posters are much easier to transport than paper posters.

24.8 PRESENTING A POSTER PRESENTATION

Check the instructions, including when and how you should hang your poster. Conference organisers may provide fixings, but it's useful to have drawing pins and self-adhesive fasteners with you, just in case. A pocket with handouts of your poster pinned near it is very helpful; you'll need to replenish these during the meeting. A sign-up sheet for viewers to add their email address if they'd like an electronic version of the poster is also useful.

Be at your poster at the scheduled time so that you can meet people and discuss your work with them. It can feel intimidating to stand by your work, but doing so will lead to interesting conversations, new contacts, and useful feedback. You may wish to practice a short introduction to your poster, to break the ice. Talk to people, be friendly and offer to explain your work to them. If friends come by, ask them not to deter other visitors from viewing your poster.

Poster sessions can be crowded, and often include food and drinks. Remember that you don't need everyone to read your poster, just those who are interested. As with oral presentations, check that you understand questions, and ask for clarification if needed. Answer questions briefly and directly, politely and courteously. Thank viewers for visiting your poster.

Many conferences invite poster presenters to give **lightning talks** to encourage people to view your poster. These are very short presentations, often just 1 or 2 minutes long, with strict timing. Lightning talks require you to make your point clearly and present only critical information. You may be allowed slides. If you are, follow the instructions carefully, as these will probably be combined into a single presentation with other presenters, which can upset your formatting. Use your presentation to advertise your poster and tell people where to find it. Provide a taster of what you have to say. There's no need to cover the entire content; you want people to visit your poster to find out more.

24.9 CONFERENCE ETIQUETTE

If you ask a question after an oral presentation, be positive and offer constructive suggestions. Save negative feedback for a one-on-one conversation. Use questions to seek information, not to showoff your own knowledge. Don't get embroiled in a debate with the speaker; other people may have questions, too. If you have a lot to discuss, talk to the speaker after the session.

At meetings, researchers share recent work which is often unpublished. Do not photograph or otherwise record people's presentations without their permission. If you don't want your own research broadcast, include a clear *do not share* logo in your presentation. It is unethical to use unpublished information you gain from a conference presentation to compete with the presenters (Section 3.1).

The networking opportunities at conferences often blur social and professional lines and encompass power differentials. This can, and does, lead to inappropriate and unacceptable behaviour, including sexual harassment and racial insults. Some conferences provide statements on appropriate behaviour and guidance on how to deal with unacceptable, unexpected, and unwanted behaviours, including abuses of power. Encourage conference organisers to do this if they do not already do so, to promote an inclusive and safe environment for all. Report any abuses to meeting organisers if you feel you can do so, and be an effective ally to victims of unacceptable behaviour. Senior primatologists have a particular responsibility here.

24.10 CHAPTER SUMMARY

In this chapter, we've seen that:

- Conferences are excellent opportunities to discover the latest findings in our field, meet other researchers, and get feedback on our work.
- Abstract deadlines may be months before the conference.
- We should attend to diversity when inviting people to contribute to a symposium.
- We should check presentation guidelines carefully and follow them.
- We should communicate clearly and concisely in presentations and avoid distracting our audience.
- We should prepare and practice presentations well in advance and incorporate feedback.
- We should tailor presentations to our audience.
- We should not be tempted to read out an oral presentation or to read our slides to our audience.
- We should reduce text to a minimum in presentations, simplify figures as far as possible and avoid tables.
- We should answer questions briefly, directly, and politely.
- We must observe conference etiquette and behave ethically.

The next chapter summarises the key points and conclusions from our journey through the process of studying primates.

24.11 FURTHER READING

Ally M. 2013. *The Craft of Scientific Presentations: Critical Steps to Succeed and Critical Errors to Avoid*. 2nd edn. New York. Describes an assertion-evidence approach to presentations. See also www.craftofscientificpresentations.com.

Calisi RM and a Working Group of Mothers in Science. 2018. Opinion: How to tackle the childcare–conference conundrum. *Proceedings of the National Academy of Sciences of the United States of America* 115: 2845–2849. https://doi

.org/10.1073/pnas.1803153115. Suggestions for how to address childcare-related barriers to conference attendance.

Hailman JP, Strier KB. 2006. *Planning, Proposing, and Presenting Science Effectively.* Cambridge: Cambridge University Press. Chapter 3 covers conference presentations.

Isbell LA, Young TP, Harcourt AH. 2012. Stag parties linger: Continued gender bias in a female-rich scientific discipline. *PLOS ONE* 7: 2–5. https://doi.org/10.1371/journal.pone.0049682. Shows gender bias in invitations to contribute to primatology symposia at the American Association of Physical Anthropology meetings. Symposia had half the number of female first authors than did submitted talks and posters. Moreover, symposia organised by men had half the number of female first authors than did symposia organised by women or by both men and women.

Karban R, Huntzonger M, Pearse IS. 2014. *How to Do Ecology: A Concise Handbook.* 2nd edn. Princeton, NJ: Princeton University. Chapter 8 includes oral presentations and posters, with checklists of suggestions for each.

Langin KM. 2017. Tell me a story! A plea for more compelling conference presentations. *The Condor* 119: 321–326. http://dx.doi.org/10.1650/CONDOR-16-209.1. Presents strategies for leading your audience through your presentation.

Matthews JR, Matthews RW. 2014. *Successful Scientific Writing: A Step-by-Step Guide for the Biological and Medical Sciences.* 4th edn. Cambridge: Cambridge University Press. Chapters 15 and 16 cover oral presentations and posters.

Rosen J. 2016. Find your voice. *Nature* 540: 157–158. http://dx.doi.org/10.1038/nj7631–157a. On networking and promoting your work for introverts.

Instructions for the conference you're attending.
Many websites give advice on conference presentations.

25

Conclusions

In this book, we've looked at how we study primates, including how we assess published studies, identify and develop a research question, formulate testable hypotheses and predictions, design and conduct a study that will test the predictions, select appropriate measures and samples, analyse the data, interpret the results, draw conclusions from the results in relation to the original question, and report the results in writing and presentations.

In this final chapter, I summarise the key points we have covered. I also provide some career advice for primatologists in Box 25.1.

1. We should conduct ethical, honest, rigorous, and transparent science.
2. We must follow the 3Rs (replacement, reduction, and refinement) when working with animals, and avoid harm to human participants, the environment, the people we work with and alongside, our discipline, and wider society.
3. Science involves working with other people, and respect and good communication are essential.
4. We should help to create an equitable global scientific community by understanding social privilege and marginalisation and actively promoting inclusive science.
5. A well-designed study requires a great deal of careful thought before we start collecting data. Poorly designed studies can be a sad waste of time and effort because they don't answer the question we wanted to address.
6. Reading the literature is key to finding out what scientists already know and informs our project design and how we interpret our findings. Critical reading takes practice. We shouldn't be

distracted by the authors' claims. Instead we should examine their reasoning and the evidence they present carefully, including the study design, reliability of their measures, and the size and direction of estimated effects in addition to the results of null hypothesis statistical testing.

7. Research questions are conceptual and are always broader than the particular case study we examine them in. This means that we must read broadly, rather than focussing on our study species.

8. In applied research, we must involve those we wish to influence from the beginning of the project.

9. Once we have defined a clear research question, we formulate hypotheses that propose possible answers to it. From there, we derive predictions from hypotheses to test their validity with empirical data.

10. Well-formulated predictions tell us exactly what data we need to collect to test them and how to analyse the outcome. In other words, predictions dictate our study design, variables, and data analysis.

11. We can control variation in one or more predictor variables to reveal the effect of each one on an outcome variable statistically or experimentally. Each has advantages and disadvantages.

12. Data analysis is an integral part of study design and we can't wait until we have our data to think about how to analyse it.

13. Study design includes careful consideration of the number of true replicates we need to test our predictions. We use prospective power analyses to make informed judgements about sample size, minimum detectable effect sizes, and the relative risk of false positives and false negatives.

14. Detailed research planning helps to determine the feasibility of a study and anticipate problems. Pilot studies are vital to test methods, troubleshoot, and find out what is and isn't feasible. They can also reveal exciting new possibilities. Detailed timelines and budgeting ensure a project is feasible. We also need to consider risk to ourselves and people we work with.

15. The stages of the research process feed back on one another and it is normal to abandon ideas and start again many times. We might go back one stage; for example we might rethink our predictions if we can't measure what we want to measure. We might go back several stages and choose a different study site, taxon, or design. We might even need to rethink our entire research question. This

can be daunting, but the research design process is easier each time we go through it.

16. Preregistration of research plans, including a detailed data analysis plan, addresses problems of *post hoc* theorising, improves science, and has great potential benefits for our field.

17. Writing winning funding proposals takes good planning, a lot of time, and feedback. We can't do it at the last minute. Funding proposals must be tailored to the funder; we can't cut and paste from one application to another.

18. Data collection rarely goes to plan, but a plan is still crucial. We must focus on our goals and monitor our progress towards them, but we must also be flexible and take opportunities that present themselves. We must be both patient and determined. Fieldwork, in particular, can be physically, psychologically, and emotionally very challenging. We must collect data rigorously and be blind to the question if possible. We must not peek at our data as we go along. We must keep data and samples safe.

19. It's very unlikely that we will collect exactly the data we set out to collect, but we must avoid the temptation to explore our data, looking for patterns, to avoid detecting and reporting false positives. We must also resist the temptation to run multiple analyses of the same dataset in the hope of finding support for a prediction. Instead we should conduct our pre-planned analyses and specify any exploratory analysis decisions made after examining the data clearly. We should test the assumptions underlying the tests we use and assess the relative risk of false positives and false negatives when considering our results. We should calculate and report effect sizes and confidence intervals as well as assessing statistical significance.

20. Disseminating our findings is part of the scientific process. We must report studies honestly. This includes describing the rationale and method for selecting our sample size and describing our data so the readers can understand it. We must report all statistical analyses and clearly distinguish between exploratory and confirmatory analyses. We must report the size and direction of all estimated effects, with confidence intervals, and report exact p values for all null hypothesis statistical tests. Sharing our data and code makes our research transparent and advances science.

Box 25.1: Career advice for primatologists

Careers in primatology are as varied as primatologists are, due to individual differences in our aims and aspirations, and regional differences in circumstances. As a result, there is no single way to get on in primatology. In all cases, however, you need to study and understand the system you work within to succeed in it.

Primatologists work in conservation organisations, animal sanctuaries, zoos, and academic institutions, among others. Many contracts in, and associated with, primatology are precarious, with short-term and part-time contracts, low wages, and a lack of job security.

The specific criteria a selection panel looks for vary with the role and the system you work within. Once you have a job, career progression may involve applying for promotion, or changing employer.

When thinking about job applications, consider your personal aims, what you enjoy most, and where your skills lie. The skills involved in research, including critical reading, designing and planning a project, assessing evidence, interpreting data, working with others, and communicating with expert and lay audiences, are all useful for a great variety of jobs and careers. Also consider practicalities; jobs may require you to move across countries and may not pay enough to cover your needs.

Jobs in primatology are advertised online, and on mailing lists. Conferences (Chapter 24) are a good source of information about upcoming opportunities, and a great way to meet potential employers.

If you know what sort of post or promotion you're aiming for, start early by assessing your experience and skills in relation to the criteria. You may not yet have the skills, experience, and track record required for a post or a promotion. If this is the case, determine what you need to strengthen, and seek opportunities to fill the gaps.

Much of the advice for funding applications (Chapter 18) applies to job applications and career progression, too. It is crucial to understand the system that you're applying to. Find out as much as you can about how the system works. This includes formal policies and informal information on unwritten practices. Consider any informal information you receive carefully; it may not be accurate.

Box 25.1: (cont.)

Seek advice from mentors and ask them for feedback on draft applications. This is particularly important if you have imposter syndrome (Box 1.4). Talk to people involved in assessing candidates and search online for advice relevant to your situation.

Like funding applications, job and promotion applications are often read (or just scanned) by very busy people. This means that you must do as much of their work for them as possible. Tailor your application to the post; don't simply send your CV with a standard cover letter. If there are stated criteria, explain, with evidence, how you fit them all. Don't simply say that you meet the criteria. Don't expect assessors to find and consider information that you have not brought to their attention. A table of the criteria with evidence for how you fit them can be useful.

Be honest, but don't under-sell yourself. This is not a time to be humble.

Check your spelling and grammar. A poorly crafted application suggests that you don't care about the opportunity.

Many applications require references; follow the advice in Box 18.1.

Potential employees, search panels, and promotion committees may search for you online, so ensure that your personal website is professional and up-to-date and includes the information they are likely to look for.

If you're called to interview for a position, find out as much as you can about the institution, the interview process, and the panel. Find out what sort of dress is appropriate and dress accordingly. Don't assume that the panel members have read your application and don't be shy about repeating information that you want them to be aware of. Be ready to explain why you want the position. This is your chance to show that you have researched the position; *I need a job* is not a good answer. You will probably be asked about your future plans, so have an appropriate response ready.

Interviews often include the opportunity to ask questions, as well as answer them, so be prepared for this. This is a good opportunity to show your interest in the job. This is often the last part of an interview, so it's a chance to make a good final impression.

As with funding applications, there are usually more excellent candidates than there are posts available. If you are shortlisted for a

Box 25.1: (cont.)

position, but unsuccessful, this means the panel selected someone else. It doesn't necessarily reflect on you or your performance. Seek feedback, and reflect on your performance to learn from it, and then move on to your next application. As with funding applications (Chapter 18) and manuscript submission (Chapter 23), rejection is part of the process. There's a fine line between thinking carefully about applications and the assessment process, and overthinking.

Time and effort invested in an application are much more likely to pay off than indiscriminate approaches. For example, if you hope to work with someone as a student or a postdoctoral researcher, read his or her work and explain why you are interested in working with him or her; don't just say you find his or her work interesting without explaining why. Proposing ideas for a collaboration is much more likely to be successful than just sending your CV and expecting the person you contact to come up with possibilities. Focus on what you can offer the person you're writing to, not what you want from them. Good luck!

FURTHER READING

Kelsky K. 2015. *The Professor Is In*. New York: Random House. Advice on graduate school, the job market, and tenure. US-focussed but very useful for researchers from other countries, too. See the author's website at http://theprofessorisin.com.

Kruger P. 2018. You are not a failed scientist. *Nature* 560: 133–134. On job prospects for PhD students and making informed decisions about careers.

Moreno E, Gutiérrez J-M. 2008. Ten simple rules for aspiring scientists in a low-income country. *PLOS Computational Biology* 4: e1000024. https://doi.org/10.1371/journal.pcbi.1000024. Proposes ten rules to help scientists from low-income countries to mature as professional scientists in their home country.

Index